Mobile Phone Panel Surveys in Developing Countries

Contents

Tables

Foreword

Poverty must be quantified before it can be defeated. Over the past two decades, considerable progress has been made in measuring the welfare of populations in developing countries. While traditional welfare surveys will continue to be the cornerstone for measuring poverty and related aspects of household welfare, these surveys are less suited for monitoring rapidly changing conditions. Yet, decision makers increasingly demand to be informed on a timely basis, particularly at times of crisis. The ability to gather real-time data on living conditions to identify, for instance, the first signs of food shortage could help monitor and assess the impact of a food crisis. It could also facilitate more effective and targeted interventions.

The proliferation of mobile phone networks and inexpensive handsets has opened new possibilities to collect data with high frequency. Initiatives such as the World Bank's *Listening to Africa* have exploited these new opportunities. Over the past couple of years, various Bank and non-Bank teams have successfully tested approaches to collect reliable and representative welfare data using mobile phone surveys. Mobile phone surveys have been fielded in a range of countries, including Tanzania, Madagascar, Togo, Malawi, and Senegal. At times, these surveys were fielded under the most challenging of circumstances, as was the case during the Ebola outbreak in Sierra Leone and the security crisis in Northern Mali. In all instances, reliable and representative data were obtained, at a reasonable cost.

This book, *Mobile Phone Panel Surveys in Developing Countries: A Practical Guide for Microdata Collection*, reflects practical lessons from these efforts. It will be of great relevance to those considering collecting representative data using mobile phones themselves. This is a new frontier in statistical development. The mobile revolution is changing the way we work. For the better.

Makhtar Diop
Vice President, Africa Region
The World Bank

Acknowledgments

This study has been conducted as a joint effort between the Poverty and Equity Global Practice–Africa Region of the World Bank and Twaweza. The handbook is authored by Andrew Dabalen (Lead Economist, Poverty and Equity Global Practice), Alvin Etang (Economist, Poverty and Equity Global Practice), Johannes Hoogeveen (Lead Economist, Poverty and Equity Global Practice), Elvis Mushi (former Research Manager, Twaweza), Youdi Schipper (Manager, KiuFunza Randomized Evaluation, Twaweza), and Johannes von Engelhardt (Erasmus University of Rotterdam). It has also benefited from contributions of Felicien Accrombessy (Senior Economist/Statistician, Poverty and Equity Global Practice), Kristen Himelein (Senior Economist/Statistician, Poverty and Equity Global Practice), Sterling Roop (International Law and Policy Institute), and Keith Weghorst (Vanderbilt University).

The authors acknowledge Kathleen Beegle and Nobuo Yoshida for their extensive comments as peer reviewers at the decision meeting chaired by Punam Chuhan-Pole (Lead Economist, Africa Regional Office of the Chief Economist). The handbook also benefited from helpful suggestions from numerous individuals outside the World Bank. In particular, acknowledgments are due to Brian Dillon (Assistant Professor, Evans School of Public Policy and Governance, University of Washington), Deborah Johnston (Head of Department, School of Oriental and African Studies, University of London), and the Twaweza team led by Aidan Eyakuze (Executive Director), particularly Melania Omengo, John Mugo, Nellin Njovu, Sana Jaffer, and Victor Rateng. The authors are grateful for the support and guidance of Pablo Fajnzylber (Practice Manager, Poverty and Equity Global Practice). Finally, the team is grateful to partners in implementing countries for sharing their experiences, especially the national statistical offices of Madagascar, Malawi, Senegal, and Togo, Human Network International, GISSE (Mali), and Ipsos Synovate (Tanzania).

The handbook is a key output of the World Bank's Listening to Africa Initiative (that is, mobile phone panel surveys), which was launched in 2012. The work program, including this study, has been financed through grants from the Statistics for Results Facility Trust Fund, the Umbrella Facility for Gender Equality Trust Fund on Gender-Informed Mobile Phone Surveys in Africa, and the BNP Paribas Trust Fund on Growth and Equity: Welfare and Services Monitoring across Sub-Saharan Africa via Mobile Phone Surveys.

About the Authors and Contributors

Andrew Dabalen is a Lead Economist in the Poverty and Equity Global Practice at the World Bank. He focuses on policy analysis and research on development issues such as poverty and social impact analysis, inequality of opportunity, program evaluation, risk and vulnerability, labor markets, and conflict and welfare outcomes. He has worked in the World Bank's Sub-Saharan Africa and Europe and Central Asia Regions. He has coauthored regional reports on equality of opportunity among children and on poverty in Africa. He has published scholarly papers on poverty measurement, conflict and welfare outcomes, and wage inequality. He holds a PhD in agricultural and resource economics from the University of California, Berkeley.

Alvin Etang is an Economist in the Poverty and Equity Global Practice at the World Bank. Previously a Postdoctoral Associate at Yale University, Alvin holds a PhD in economics from the University of Otago, New Zealand. His interest in microdevelopment has led him to focus on the analysis of poverty and welfare outcomes. With substantial experience in household survey design and implementation, Alvin has worked in several African countries. He is currently managing the World Bank's Listening to Africa Initiative on mobile phone panel surveys for welfare monitoring. He has also taught economics at the undergraduate level and has experience in designing and using economic experiments as a tool to analyze poverty issues. His research has been published in peer-reviewed academic journals and has featured in the popular press such as *The Economist, The Wall Street Journal, Financial Times, The Atlantic,* and *Frontline.*

Johannes Hoogeveen is a Lead Economist in the World Bank's Poverty and Equity Global Practice. Based in Washington, he coordinates much of the practice's program for francophone Africa. He combines analytical work with operational work: budget support as well as investment lending. His latest project is a regional program to harmonize and modernize multimodular living conditions surveys in the eight member states of the West African Economic and Monetary Union. His current research interest revolves around enhancing development effectiveness in poor governance environments. He was manager at Twaweza in Tanzania, where he led the initial efforts to design functional mobile phone panel surveys. He holds a PhD in economics from the Free University Amsterdam.

Elvis Mushi is the Head of Research at the Financial Sector Deepening Trust in Tanzania, where he leads in the design and implementation of studies related to the financial sector. With the desire to encourage evidence-based decision making in the financial sector, he has pioneered an ambitious data-to-solution initiative, dubbed FinDisrupt, that brings together market players to use data and insights in developing product and policy solutions. Prior to his current tenure, he served as Regional Manager for Sauti za Wananchi, Africa's first nationally representative mobile phone panel survey, in Tanzania. He played a crucial role in designing, testing, and rolling out of the survey in Tanzania and in its expansion to Kenya. He holds a BSc (Hons) degree in international business administration with a double concentration in finance and marketing from United States International University in Nairobi.

Youdi Schipper is Research Advisor at Twaweza East Africa and senior researcher at the Amsterdam Institute for International Development. His work focuses on education reform, impact evaluation, and mobile data collection. He has initiated and implemented Sauti za Wananchi in Tanzania, findings from which regularly feature in the East African media. He is managing KiuFunza, a nationally representative randomized trial studying the impact of teacher performance pay in public primary schools in Tanzania. He is contributing to the Research on Improving Systems of Education study on education reform in Tanzania. He holds a PhD in economics from Free University Amsterdam.

Johannes von Engelhardt is an Independent Research Consultant at the World Bank. Based in Amsterdam, he has worked on mobile panel surveys for the World Bank and other development organizations in Gaza, Senegal, Tanzania, and Togo. He holds a research master's degree in the social sciences from the University of Amsterdam and is currently finishing a PhD in media and communication at Erasmus University Rotterdam, where he also works as a Lecturer.

Félicien Donat Edgar T. Accrombessy is a Senior Economist and Statistician at the Africa Region of the World Bank and Poverty Economist for Benin. He has more than 18 years of experience in designing household surveys and measuring and analyzing poverty. For several years, he worked in statistical capacity building, monitoring, and evaluation in development programs and projects and on the coordination, harmonization, and implementation of statistical information systems, indicators, and databases, including statistics master plans and statistics development strategies. Before joining the World Bank, he worked as Chief of the Department of Informatics and Statistics at the National Agency for Employment, Benin, where his core role was managing the database on job seekers and producing studies on employment and key labor market indicators. He also worked five years at the Benin National Institute for Statistics, where he was in charge of statistical coordination. He holds a master's degree in economic analysis and international development and a master's degree in public decision and development project management.

Kristen Himelein is a Senior Economist and Statistician in the Poverty and Equity Global Practice at the World Bank. Her areas of expertise are poverty analysis, scientific sampling, and survey methodology. She specializes in nontraditional sample designs in challenging developing-world contexts and has ongoing work in Chad, Liberia, Sierra Leone, and Somalia. She holds a master's degree in public administration in international development from the Harvard Kennedy School and a graduate degree in survey statistics from the Joint Program on Survey Methodology at the University of Maryland.

Sterling Roop is a Senior Adviser at International Law and Policy Institute, where he works on politics, accountability, and conflict in East Africa and in the Horn of Africa. His research interests include political economy, legal sector reform, human rights, and governance. He recently led the Tanzania Towards 2015 project for International Law and Policy Institute investigating the social, political, and economic drivers of fragility in Tanzania and in Zanzibar using both qualitative and quantitative survey methods. He has a master's degree in international relations from Boston University.

Keith Weghorst is a Political Scientist at Vanderbilt University. He is currently a postdoctoral research fellow and will become assistant professor there in August 2016. His research focuses on elections in Africa, with a specific interest in political parties, legislative candidacy, and women's representation in the subcontinent. He is currently working on a book on opposition candidacy in electoral authoritarian regimes. The book draws upon several years of dissertation fieldwork in Tanzania. In addition to his substantive focus, he specializes in survey methodology and techniques to improve the quality of citizen and elite survey data collected in developing-country settings. His work has been published in the *American Journal of Political Science*, *Comparative Political Studies*, and *Democratization*. He holds a PhD in political science from the University of Florida and a master's degree in African studies from the University of California, Los Angeles.

Abbreviations

CAPI	computer-assisted personal interviewing
CATI	computer-assisted telephone interviewing
EA	enumeration area
IVR	interactive voice response
LSMS	Living Standards Measurement Study
MPPS	mobile phone panel survey
PAPI	paper-and-pencil interviewing
RDD	random digit dialing
SIM	subscriber identity module
SMS	short message service
USSD	unstructured supplementary service data
WAP	wireless application protocol

Introduction

Household survey data are useful for monitoring the living conditions of citizens in any country. In developing countries, a lot of these data are collected through traditional face-to-face household surveys. Because of the remote and dispersed nature of many populations in developing countries, but also because of the complex nature of many survey questionnaires, the collection of timely welfare data has often proven expensive and logistically challenging. At the same time, there is a need for more rapid, less expensive, less complicated, and more nimble data collection methods to address the data gaps between big household surveys.

During the past decade, there have been promising new opportunities, backed up by some empirical data, for the collection of statistics at much higher frequency. The recent proliferation of mobile phone networks and inexpensive telephone handsets has opened new possibilities. Based on a combination of baseline data from a traditional household survey and subsequent interviews of selected respondents using mobile phones, a growing number of initiatives are now using mobile phone technology to facilitate welfare monitoring and opinion polling almost in real time.

The purpose of this collection of seven chapters is to contribute to the development of the new field of mobile phone data collection in developing countries. This handbook documents how this innovative approach to data collection works, the specific advantages of mobile phone panel surveys (MPPSs), the practical aspects, and the challenges in survey design. The handbook mainly draws on the first-hand experiences of the authors with a number of mobile data collection initiatives, primarily across Sub-Saharan Africa. In addition, the text builds on the experience and lessons gathered through mobile data collection projects in other regions of the world.

The mobile phone surveys discussed in this handbook are particularly relevant for initiatives endeavoring to collect limited sets of representative data in an affordable manner. The approach described represents an indispensable tool for individuals, organizations, and others seeking feedback from populations, monitoring changes in living conditions over time, or analyzing the impact of interventions.

The handbook is intended to serve a diverse audience, including people involved in collecting representative data using mobile phones (for example, statistical offices, researchers, and survey firms) and people using data collected through the mobile phone approach (for instance, governments and policy makers, donors, and international organizations). The chapters can be used to guide individuals implementing an MPPS through every stage of the implementation process. In this sense, the handbook can be read as a cookbook for representative MPPSs. For potential users of the data collected through mobile phone technology, the chapters present a new approach to data collection that can be applied in monitoring programs and to facilitate nearly real-time decision making. Furthermore, the contents of this handbook equip the audience with background knowledge to assess the quality of the data they use in decision making. An additional purpose is to contribute to the debate on the advantages of the method as well as the challenges associated with it.

The seven chapters are organized as follows. First, in chapter 1, the rationale for using MPPSs is articulated, and the key issues are highlighted. Chapter 2 provides a concise overview of the main issues that should be considered in the design of a representative mobile phone survey. Chapter 3 focuses on baseline survey preparation and implementation. The chapter presents the reader with a set of questions that need to be answered before carrying out fieldwork for the baseline survey. The chapter also takes the reader step-by-step through the fieldwork involved in implementing a large-scale, nationally representative MPPS baseline survey. The next two chapters cover phone survey round preparations and implementation. The decisions and steps to be taken in setting up a call center for an MPPS are described in chapter 4, while chapter 5 discusses elements of a typical mobile phone survey round, highlighting those aspects that are specific to mobile phone panels. Chapter 6 discusses four important components that follow on data collection: data analysis, report writing, dissemination of results, and making data available in a public forum. After going through these chapters, the reader should have acquired the knowledge necessary to implement an MPPS. However, in the end, the decision to implement an MPPS will depend to a large extent on the cost of conducting the survey. Project budget and cost are always among the issues raised when the MPPS approach is shared with colleagues. The final chapter provides a synopsis of key budget items in an MPPS project, with examples of the budgets in actual projects.

CHAPTER 1

The Rationale for Mobile Phone Panel Surveys

Why Mobile Phone Panel Surveys?

Development agencies, policy makers, and analysts working in and on developing countries are in constant need of data that are reliable, relevant to context, timely, gathered regularly, and comparable over time. High-quality data on the living conditions, behavior, and opinions of populations are needed for diverse purposes, such as monitoring poverty, demographic changes, public opinion, beneficiary feedback, and labor market participation; research; capturing the impact of shocks, development programs, and policies; and tracking government program implementation and private sector development.

In developing countries, much of these data are gathered through traditional face-to-face household surveys. Because other sources of administrative data are often incomplete, less reliable, or dated, household surveys are a key resource for knowledge on economic and social realities. Indeed, "in developing countries, [household surveys] have become a dominant form of data collection, supplementing or sometimes even replacing other data collection programs and civil registration systems" (United Nations 2005, iii). A number of large-scale household survey programs are conducted in almost all developing countries with some regularity.[1]

At the same time, despite the advances in computer-assisted personal interviewing (CAPI) technologies, household surveys tend to be costly because of their complexity. Designing and implementing a survey and producing a clean dataset ready for analysis can sometimes take several months. Together, these factors may lead to significant delays between the formulation of survey questions and the availability of data, often with turnaround times of at least a year. While traditional household surveys provide invaluable insights into people's everyday lives, the high costs, low frequency, and long turnaround times mean that they often fail to meet urgent data needs.

This situation is particularly problematic, because—as articulated in recent research on the worrisome state of African statistics (Jerven 2013;

Jerven and Johnston 2015)—outdated and unreliable basic statistics are a major obstacle to formulating and evaluating economic policy in developing countries. Reflecting on the quality of data on growth and poverty in Africa, Devarajan (2011) writes of a "statistical tragedy."

The last 10 years have seen a surge of interest in the use of new technologies to gather high-quality, high-frequency survey data on the living conditions and perceptions of populations in developing countries. With the availability of inexpensive phone handsets and rapidly growing network coverage in many developing countries, the mobile phone has attracted much of the attention as a new tool for collecting high-frequency and, oftentimes, low-cost survey data.

Utilized to track food security in refugee camps (WFP 2015), run nationally representative multipurpose citizen panels (for example, the World Bank's Listening to Africa Initiative and the Sauti za Wananchi [Voices of Citizens] survey in Kenya and Tanzania), monitor the harvest expectations of farmers (Dillon 2012), or track changes in welfare (Etang, Hoogeveen, and Lendorfer 2015; Himelein 2014), mobile phone surveys have recently been employed by various organizations and researchers in data gathering efforts in developing countries.[2]

In light of the proliferation of mobile phone surveys, this handbook seeks to provide guidelines and describe best practices for this relatively new form of data collection. It thereby aims to contribute to the available resources documenting the experiences, problems, and solutions in conducting mobile phone panel surveys (MPPSs) in developing countries.

The MPPS Approach

In this handbook, an MPPS refers to a process for collecting *representative high-frequency panel data in developing countries by using mobile phones*. The focus is on representativeness: the MPPS approach has roots in data collection methods based on representative household surveys. The MPPS is different from approaches such as web-polling or the U-Report system, which cannot guarantee the representativeness of the samples. The spotlight is on developing countries because, in these settings, phone-based interviews are still a novelty. In developed countries, phone interviews are common, and there is a large literature discussing the pros and cons of the relevant approaches (see Murgan 2015).[3] The MPPS approach is centered on collecting information repeatedly from the same respondent (panel data) because this has proved to be the most cost-effective way to implement a mobile phone survey in a developing-country context.

This handbook presents a two-step approach: a face-to-face baseline survey combined with follow-up survey rounds conducted through mobile phones. Baseline respondents who do not own a mobile phone can be given a phone during the face-to-face interview, and small incentives in the form of airtime credits are transferred after each completed mobile phone interview round (see chapter 3).[4] The approach was piloted in a few developing settings, including Dar es Salaam, Tanzania; Guatemala; and urban South Sudan (Demombynes, Gubbins, and Romeo 2013; Hoogeveen et al. 2014;

Schuster and Perez-Brito 2011), Twaweza initiated the Dar es Salaam survey and used the findings to scale up the survey model across mainland Tanzania.[5] The International Law and Policy Institute is implementing the same approach in Zanzibar, where it is referred to as Wasemavyo Wazanzibari Mobile Survey.[6] MPPSs are also being carried out in several countries, including Madagascar, Malawi, Mali, Senegal, and Togo, as part of the broader World Bank regional Listening to Africa Initiative.[7]

The succeeding chapters will go over the practicalities, advantages, and challenges of the MPPS design, drawing primarily from the firsthand experiences of the authors, which cover MPPSs in a number of African countries, particularly Madagascar, Malawi, Mali, Senegal, Sierra Leone, Tanzania, and Togo, that were implemented by various organizations such as the World Bank and Twaweza. In addition, the handbook draws on available documentation on lessons learned and personal conversations to integrate experiences with mobile phone data gathering projects implemented by other organizations, such as the World Food Programme.

Before the more detailed discussion of the potential utility of this novel form of data collection and some of the issues in implementation, two points should be stressed. First, the data collection approach proposed in this handbook does not replace traditional data collection approaches (including household surveys, censuses, and administrative data collection). The approach is complementary. Mobile phone surveys are not suited for lengthy interviews and are rather more effective for monitoring rapidly changing conditions and obtaining real-time feedback from households.

Second, most of the examples provided in this handbook are drawn from existing mobile phone surveys in Africa, where the authors mostly work. However, this approach is useful in many other developing countries. Similar survey approaches have been tested in Guatemala, Honduras, and Peru and mobile phone surveys have been piloted in Bangladesh and Serbia to obtain quick poverty estimates (Ballivian et al. 2013; Boznic et al. 2015; Schuster and Perez-Brito 2011).

Advantages of MPPSs

Mobile phone surveys should not and cannot replace face-to-face household surveys in developing countries. However, in specific circumstances and for specific data needs, mobile phone surveys offer substantial benefits. In this section, we provide an overview of some of the main advantages.

Gathering Data in Volatile and High-Risk Environments

The added value of mobile phone surveys often becomes most evident in situations where face-to-face data collection would be extremely difficult or simply not feasible. For example, the West Africa Ebola virus outbreak in 2014 limited the movement of people and goods in the three most affected countries (Guinea, Liberia, and Sierra Leone), all but halted international trade and tourism, disrupted payment systems, raised prices, reduced the availability of essential goods, and disturbed agricultural production. While there was an urgent need to monitor these impacts, the health and security situation did not permit the deployment of enumerators. Mobile phone surveys were used to collect data to monitor the Ebola crisis and its effects on food security and to provide estimates of its socioeconomic toll (Himelein 2014; WFP 2015).

MPPSs have also been used by the World Food Programme to gather data about food security among internally displaced persons in refugee camps in eastern Democratic Republic of Congo and Somalia. Similarly, the World Bank's Listening to Displaced People Survey collected information on living conditions from the displaced population (internally displaced persons and returnees) in Mali and refugees in camps in Mauritania and Niger who had been displaced by the crisis in north Mali (Etang, Hoogeveen, and Lendorfer 2015). These examples illustrate how mobile phone surveys are used to obtain information from populations that would otherwise be difficult or impossible to reach.

Collecting Data in Remote Areas and Enumerator Safety

Another advantage of mobile phone surveys is that they facilitate the survey of households in remote areas. Enumerator safety seems to be another advantage of this data collection approach in areas where it is difficult to travel at night after interviewing people in the evening.

Quick Response to New Data Needs

Typically, mobile phone surveys offer a high degree of flexibility and short turnaround times, allowing implementers to react quickly to new and unexpected data needs. For example, when Malawi was affected by floods in January 2015, the Listening to Malawi survey team was able to respond rapidly, running a short survey within a couple of days to collect information on the severity of the crisis and the impact on households. Producing this type of real-time data about a situation can facilitate timely and more effective intervention. Timeliness is a key strength of the MPPS approach.

Data Quality

With the proliferation of mobile phone interviewing, there is now also growing evidence on the quality of data produced by this mode of surveying. Lynn and Kaminska (2013) use a randomized experiment in Hungary to investigate whether mobile phone survey data differ from data collected using landlines. The authors start by identifying four reasons why the quality of data collected using mobile and fixed phone interviews might differ: line quality, the extent of multitasking among survey respondents, the extent to which survey respondents are distracted from the task of answering questions, and the extent to which other people are present and able to overhear what the survey respondent is saying. They evaluate the effect of differences in these features on survey measures through an experiment in which a sample of people who have both mobile and fixed phones are randomly assigned to be interviewed either on their mobile phones or on their fixed phones. They find only small variations in survey measures between mobile phone interviews and fixed phone interviews and suggest that data quality may be higher in mobile phone interviews. A more recent study confirms that data collected using mobile phones are high quality. Garlick, Orkin, and Quinn (2015) conduct a randomized controlled trial to test whether high-frequency data collected from microenterprise owners through interviews over the phone in South Africa are useful and viable. They find no differences in data collected during weekly face-to-face interviews and data collected during weekly mobile phone interviews. Similarly, a World Bank mobile phone survey study in Honduras carried out by Gallup found no significant differences between responses collected through face-to-face interviews and through mobile phone interviews (Ballivian et al. 2013).

The fact that others cannot overhear the questions being asked may sometimes allow respondents to feel more secure in providing information over the phone rather than face-to-face. So, if sensitive information is being collected, mobile phone surveys may be preferable. A related hypothesis is that panel respondents may provide more truthful answers to call center interviewers after a relation of trust has been established.

In sum, the handbook is not propagating a claim that the data collected through mobile phones are better than data collected through face-to-face surveys. Yet, there is no evidence that the data collected through mobile phones are worse in any way. Research is needed to supply empirical evidence on which of the two approaches produces better data.

Cost-Effectiveness

Depending on the context and aim of the data gathering effort, mobile phone surveys can be cost-effective. Whether this is the case depends on the length of the questionnaire and the basis of comparison. For example, if the cost of a baseline survey and other operational costs are included, mobile phone surveys are not necessarily less expensive than traditional face-to-face interviews in the price per question. However, aside from the initial costs (the baseline survey, equipment, utilities, and setting up the call center), the low marginal cost means that a mobile phone survey can be cost-effective. If the follow-up phone surveys are extensive, the overall costs become substantially lower. Furthermore, if there are only a few questions to ask (as is often the case in monitoring, for example), an MPPS is cost-effective on a per question basis. (Chapter 7 discusses these considerations in detail and also provides the budgets of previous MPPS projects.)

Advantages in Monitoring and Impact Evaluation Efforts

There is growing demand for up-to-date information on living conditions. Decision makers need timely data to monitor the situation in their country so as to identify, for instance, a looming food crisis or signs of an emerging economic disaster.

High-frequency MPPSs can generate routine monitoring data such as price data on the main food staples, rainfall, transport costs, security, productivity during an agricultural season, and the availability of agricultural inputs such as fertilizers on markets. The surveys can also facilitate the near real-time collection of monitoring information on emergency situations such as health shocks and generate project-specific information on, for example, the targeting of voucher schemes or the disbursement of school grants. More frequent data collection is also useful in monitoring the progress toward the achievement of established goals such as the World Bank's twin goals of reducing poverty and boosting shared prosperity or the new Sustainable Development Goals.

The availability of high-frequency mobile phone panel data also offers benefits in the assessment of causality within impact evaluations. The impact of an intervention is more readily identifiable if data are collected before and after a treatment. Data collected at high frequency promotes the more accurate observation of behavioral changes as a result of a treatment. Mobile phone surveys allow this to be accomplished in a cost-effective manner. McKenzie (2012) demonstrates this formally when he shows that, if outcome measures are relatively noisy and weakly autocorrelated, as in the case of business profits, household incomes and expenditures, and episodic health outcomes, impact evaluations that rely on smaller samples and multiple follow-ups are more efficient than the prototypical baseline and follow-up model.

Any approach has strengths and limitations. Mobile phone surveys are no exception. One limitation is the restriction on the number of questions that can be asked over a phone. However, the panel nature of the surveys means that breaking down a long questionnaire into smaller parts that can be completed over the phone is possible, thereby approximating the results of a full questionnaire completed during a single face-to-face interview. The risk of attrition may also be higher in the MPPS approach than in face-to-face surveys given that the former represents a panel survey, while the latter is typically a cross-sectional survey. Across its pages, the handbook proposes ways to lessen the risk of attrition.

Inquiry on Mobile Data Collection: Voice and the Alternatives

The subsequent chapters primarily deal with computer-assisted telephone interviewing (CATI) using mobile phones. However, besides the call center–based approach that is the focus of this handbook, there are a number of other mobile phone–based modes of data collection that are worth a mention, including interactive voice response (IVR), unstructured supplementary service data (USSD), short message service (SMS), and the wireless application protocol (WAP) (see box 1.1).[8]

Each of these methods of data gathering has specific benefits and limitations. Despite the expanded role of mobile phones in data collection, few systematic studies compare the data quality and data collection efficiency of these alternative mobile approaches. A study in India has tested the accuracy of data collection using electronic forms, voice, and SMS through mobile phones (Patnaik, Brunskill, and Thies 2009). The researchers found that a voice interface in a setting in which respondents dictate answers through mobile phones to interviewers in real time

---

Box 1.1 Mobile Phone–Based Modes of Data Collection

Commonly used mobile phone–based modes of data collection include the following:

1. Voice (live interviews): computer-assisted telephone interviewing (CATI)
2. Interactive voice response (IVR)
3. Unstructured supplementary service data (USSD)
4. Short message service (SMS)
5. Wireless application protocol (WAP)

---

provides the highest level of accuracy, leading them to favor the voice approach for the implementation of a tuberculosis program. Similarly, Ballivian et al. (2013) tested various mobile data collection methods (live interviews, IVR, and SMS) in Honduras and Peru. In both countries, CATI surveys generated the lowest attrition. Furthermore, answers collected through CATI were almost identical to answers collected face-to-face, and no survey item showed a statistically significant difference.

Besides issues of attrition and data quality, technical and logistical considerations can also favor certain data collection modes over others. Hoogeveen et al. (2014) tested the effect of alternative mobile data collection methods in Dar es Salaam. They assigned households to one of four technologies—IVR, USSD, WAP, or CATI—over mobile phones. During the mobile phone panel, there were numerous problems with the different technologies. The share of respondents owning Internet-enabled phones turned out to be low (eliminating WAP); the support of the phone company to run USSD was minimal, especially once mobile banking started to claim the available bandwidth; and IVR proved clumsy because questions had to be broken down extensively to avoid too many response options. Voice did not have any of these drawbacks, nor did it necessitate that respondents be literate. Hence, after a relatively short period of time, live interviews became the technology of choice, and all people who were reachable and had access to a mobile phone were surveyed through a basic call center, which consisted of a group of interviewers who each had multiple phones (one for each phone network, allowing cheaper within-network calls) and a computer with a standard data entry screen.

This is not to say that alternatives such as IVR and SMS cannot be used for large-scale panel data collection efforts. For example, the Word Food Programme achieved promising results with IVR interviews to gather food security panel data in central Somalia (WFP 2015). While response rates were lower compared with CATI, the project managers were able to run automated IVR survey rounds successfully by paying close attention to making the questions suitable for IVR, fitting the IVR menu to local contexts, and spending sufficient time to prep respondents.

The reasons for concentrating on CATI in this handbook are the experiences of the authors in running call center–based mobile survey projects, as well as the advantages this approach offers. Live interviews allow for more flexibility in conducting interviews in different languages, posing more complex questions that may require explanations, and accommodating illiterate respondents and respondents owning low-end phones without Internet connectivity. Calling respondents on their phones can also avoid costs for the respondents. Moreover, phone interviewers can build rapport with respondents during their conversations, thus increasing trust and goodwill, reducing drop-out rates and encouraging more truthful answers, especially if question topics are sensitive, for instance, covering political preferences or opinions about local leaders.

Steps in Conducting Representative MPPSs

Combining Face-to-Face Baseline Surveys with Mobile Phone Interviews

The type of MPPS that this handbook describes comprises three steps, which are outlined here and discussed in detail in the chapters that follow: (1) a face-to-face baseline survey, (2) mobile phone interviews through a call center, and (3) data analysis, archiving, and dissemination.

This is not the only way to implement an MPPS. The handbook thus assumes that a project combines face-to-face interviews during a baseline survey with follow-up mobile phone interviews. In contrast, other approaches such as random digit dialing (RDD) exclusively rely on mobile phone interviews to call potential respondents and assess their core characteristics (location, gender, age, educational attainment, wealth, and so on). However, as detailed in chapter 2, RDD necessitates high mobile phone penetration rates and the availability of a database of telephone numbers to draw a representative sample.

Different approaches can have specific advantages in different contexts. However, given the current penetration rates of mobile phones in many developing countries, particularly in rural areas and among poor households, and data needs that can only be met through an extensive baseline survey, the approach described in this handbook is often the preferable one.

The three main steps in conducting representative MPPSs are as follows.

Step 1: Conduct a Baseline Survey

An MPPS starts with face-to-face interviews during which household and respondent characteristics are collected. The household survey needs to comprise household correlates that are likely to be of use during reporting on the mobile phone surveys. These are typically descriptive of the household and the respondent, including assets, gender, educational attainment, sources of income (including the main crops grown and any home enterprises), age, family composition, and housing characteristics. The baseline survey gathers information that is likely to be collected as part of the mobile surveys as well. The baseline survey is also the occasion to distribute free phones and solar-powered chargers in locations with limited access to electricity.[9] Chapter 3 offers guidelines for purchasing and distributing mobile phones and solar chargers.

Step 2: Implement Mobile Phone Interviews

Once the baseline has been completed and all respondents have phones, the next step is to undertake mobile phone interviews through a call center to collect household information and ask additional questions on a regular basis. Chapters 4 and 5 detail relevant considerations and best practices in the planning and implementation of the mobile phone survey rounds, including the running of a call center. A role of a call center can be as simple as using a mobile phone to call respondents and entering responses on an Excel spreadsheet or a more sophisticated data entry system.[10] Chapter 4 discusses these options.

The types of questions asked in each round will depend on the purpose and design of the survey project. While one of the advantages of mobile phone surveys is the ability to respond quickly to new data needs, asking the same questions each round is often desirable for monitoring purposes. Chapter 5 provides an in-depth discussion of the process of questionnaire design and the types of suitable questions.

Step 3: Rapid Data Analysis and Dissemination

In the analysis, data from the mobile phone survey round can be combined with data collected during the baseline survey or during earlier mobile survey rounds to allow reporting on an array of issues. Because questionnaires are typically short, analysis can be completed quickly after data gathering, and reports can be prepared and distributed shortly after data gathering. As discussed in chapter 6, data should be archived, anonymized, and made publicly available through various channels, following open data principles.

An MPPS may not be nationally representative in countries where mobile phone network coverage is far from 100 percent. However, blend approaches are feasible. For example, in the Mali mobile phone survey (see above), the refugee camp in Mauritania was surveyed by a field-based enumerator because of the weak mobile phone network. This made for a more complex survey design, but areas without network could be covered. Such a solution does not have to be expensive. It is up to the researchers to decide whether they are ready to sacrifice representativeness for ease of data collection. The baseline survey can also help investigate the extent of the difference across the households in the non-networked areas.

Although some of the surveys described in the handbook have been designed to be nationally representative, the MPPS approach proposed here is not limited to nationally representative surveys. It may be applied to subsections of a country, depending on the objective of the survey; for example, the Togo mobile phone survey targeted Lomé, the capital, only. The point is that the sample should be representative of the target population.

Notes

1. Prominent examples of large-scale household surveys in developing countries include the Living Standards Measurement Study (LSMS) surveys, the Demographic and Health Surveys, and the Multiple Indicator Cluster Surveys. The LSMS handbook provides details on the design of these types of household surveys (see Grosh and Glewwe 2000).

2. Mobile phone technology is also being used in other types of data collection that are not, strictly speaking, surveys. For example, U-REPORT, with over 200,000 participants in Uganda who respond to questions sent by short message service (SMS), is an approach that seeks feedback from volunteers. (See "U-Report," U-Report Uganda, http://ureport.ug.) Several other polling studies seek the opinions of the public across the world. The model presented in this book is distinguished from these other approaches because the data collected are intended to be representative of the target population rather than to sample people who self-select to be respondents, that is, crowd sourcing.

3. In developed countries, phone surveys (typically landline phones) and representative panel phone survey designs have been applied for some time. For an example, see endnote 1 in chapter 2.

4. MPPSs, which give phones to respondents and incentivize respondents with call credit after the completion of an interview, are, by nature, an intervention as well as a means of data collection. Especially among respondents who did not own a phone prior to participation in the survey, their ability to access information and to connect with others changes significantly. Moreover, Zwane et al. (2011) supply evidence that simply surveying respondents can lead to alterations in their responses. Furthermore, the receipt of mobile phones may also modify the status of schools in indexes that measure assets, among which mobile phone ownership is commonly included.

5. Twaweza is an East African citizen-centered open data and accountability initiative. See "Sauti za Wananchi," Twaweza, Dar es Salaam, Tanzania, http://twaweza.org/go /sauti-za-wananchi-english/.

6. For more detail, see Keith R. Weghorst and Sterling Roop, "Launch of the Wasemavyo Wazanzibari Mobile Survey," International Law and Policy Institute, Oslo, http://ilpi .org/events/launch-of-the-wasemavyo-wazanzibari-mobile-survey/.

7. See "Listening to Africa," World Bank, Washington, DC, http://www.worldbank.org /en/programs/listening-to-africa.

8. Smith et al. (2011) provide an overview of different ways to gather data using mobile phones. With IVR, respondents are called automatically on their mobile phones. Once respondents answer, a computer plays a prerecorded welcome message and proceeds to guide respondents through a menu that leads to the actual questionnaire. Respondents enter their answers by pressing keys on the mobile phone's keypad. The USSD approach allows the direct transmission of questions from a phone company server or computer to the respondent's phone; this technology also works on loaned phones. Unlike SMS messages, USSD messages create a real-time connection during a USSD session. The connection remains open, allowing a two-way exchange of a sequence of data. WAP is a web-based mobile phone survey, suited for high-end phones with Internet capability.

9. Alternatively, a village kiosk owner who provides phone charging services could be contracted to offer free phone charging to the participating households.

10. A well-set-up call center, a facility used to call and compile responses from participating households, is crucial for the success of the mobile phone survey. Management of the call center is not only labor intensive, but also requires a lot of flexibility because some households prefer to be called late in the evenings or during the weekends. As discussed in chapter 4, a call center can be run by the statistical office or can be outsourced. At the minimum, a call center comprises workstations for each of the interviewers, complete with (mobile) phones and a computer with a relevant data entry program.

References

Ballivian, Amparo, João Pedro Azevedo, William Durbin, Jesus Rios, Johanna Godoy, Christina Borisova, and Sabine Mireille Ntsama. 2013. "Listening to LAC: Using Mobile Phones for High Frequency Data Collection." Final report (February), World Bank, Washington, DC.

Boznic, Vladan, Roy Katayama, Rodrigo Munoz, Shinya Takamatsu, and Nobuo Yoshida. 2015. "Prospects of Estimating Poverty with Phone Surveys: Experimental Results from Serbia." Working paper, World Bank, Washington, DC.

Demombynes, Gabriel, Paul Gubbins, and Alessandro Romeo. 2013. "Challenges and Opportunities of Mobile Phone-Based Data Collection: Evidence from South Sudan." Policy Research Working Paper 6321, World Bank, Washington, DC.

Devarajan, Shanta. 2011. "Africa's Statistical Tragedy." *Africa Can End Poverty* (blog), October 6. http://blogs.worldbank.org/africacan/africa-s-statistical-tragedy.

Dillon, Brian. 2012. "Using Mobile Phones to Collect Panel Data in Developing Countries." *Journal of International Development* 24 (4): 518–27.

Etang, Alvin, Johannes Hoogeveen, and Julia Lendorfer. 2015. "Socioeconomic Impact of the Crisis in North Mali on Displaced People." Policy Research Working Paper 7253, World Bank, Washington, DC.

Garlick, Rob, Kate Orkin, and Simon Quinn. 2015. "Call Me Maybe: Experimental Evidence on Using Mobile Phones to Survey African Microenterprises." Working paper, Centre for Economic Policy Research and U.K. Department for International Development, London.

Grosh, Margaret, and Paul Glewwe, eds. 2000. *Designing Household Survey Questionnaires for Developing Countries: Lessons from 15 Years of the Living Standards Measurement Study*. 3 vols. Washington, DC: World Bank.

Himelein, Kristen. 2014. "Weight Calculations for Panel Surveys with Subsampling and Split-Off Tracking." *Statistics and Public Policy* 1 (1): 40–45.

Hoogeveen, Johannes, Kevin Croke, Andrew Dabalen, Gabriel Demombynes, and Marcelo Giugale. 2014. "Collecting High Frequency Panel Data in Africa Using Mobile Phone Interviews." *Canadian Journal of Development Studies* 35 (1): 186–207.

Jerven, Morten. 2013. *Poor Numbers: How We Are Misled by African Development Statistics and What to Do about It*. Cornell Studies in Political Economy Series. Ithaca, NY: Cornell University Press.

Jerven, Morten, and Deborah Johnston. 2015. "Statistical Tragedy in Africa?: Evaluating the Data Base for African Economic Development." *Journal of Development Studies* 51 (2): 111–15.

Lynn, Peter, and Olena Kaminska. 2013. "The Impact of Mobile Phones on Survey Measurement Error." *Public Opinion Quarterly* 77 (2): 586–605.

McKenzie, David. 2012. "Beyond Baseline and Follow-Up: The Case for More T in Experiments." *Journal of Development Economics* 99 (2): 210–21.

Murgan, Murtala Ganiyu. 2015. "A Critical Analysis of the Techniques for Data Gathering in Legal Research." *Journal of Social Sciences and Humanities* 1 (3): 266–74.

Patnaik, Somani, Emma Brunskill, and William Thies. 2009. "Evaluating the Accuracy of Data Collection on Mobile Phones: A Study of Forms, SMS, and Voice." Paper presented at the Institute of Electrical and Electronics Engineers' International Conference on Information and Communication Technologies and Development, Doha, Qatar, April 17–19.

Schuster, Christian, and Carlos Perez-Brito. 2011. "Cutting Costs, Boosting Quality, and Collecting Data Real-Time: Lessons from a Cell Phone–Based Beneficiary Survey to Strengthen Guatemala's Conditional Cash Transfer Program." *en breve* 166 (February), World Bank, Washington, DC.

Smith, Gabrielle, Ian MacAuslan, Saul Butters, and Mathieu Trommé. 2011. *New Technologies in Cash Transfer Programming and Humanitarian Assistance.* Oxford, United Kingdom: Cash Learning Partnership.

United Nations. 2005. *United Nations Household Sample Surveys in Developing and Transition Countries.* Document ST/ESA/STAT/SER.F/96, Studies in Methods, Series F 96. New York: Statistics Division, Department of Economic and Social Affairs, United Nations.

WFP (World Food Programme). 2015. "Post Harvest Improvement in Household Food Security Halts." mVAM Food Security Monitoring Bulletin 7, DR Congo, Mugunga 3 Camp (February), WFP, Rome.

Zwane, Alix Peterson, Jonathan Zinman, Eric Van Dusen, William Pariente, Clair Null, Edward Miguel, Michael Kremer, Dean S. Karlan, Richard Hornbeck, Xavier Giné, Esther Duflo, Florencia Devoto, Bruno Crepon, and Abhijit Banerjee. 2011. "Being Surveyed Can Change Behavior and Related Parameter Estimates." *Proceedings of the National Academy of Sciences of the United States of America* 108 (5): 1821–26.

CHAPTER 2

Designing a Mobile Phone Panel Survey

Introduction

All surveys should be launched with carefully defined objectives. Key questions to be answered include: What is the target population to be studied? What variables need to be collected? How soon and how often is the information needed? What data are already available, and how can they best be used? What are the domains of analysis on which the survey will need to present results? Will a cross-section or repeated cross-section be sufficient, or will panel data be necessary?

Among the many options available to analysts, mobile phone panel surveys (MPPSs) are suited to certain types of circumstances. As highlighted in chapter 1, an MPPS may be the best option if the required frequency is high, if relatively few variables will be collected on each round, if data are required from places where face-to-face data collection is not possible, or if the respondents are mobile and the resources needed to track them are insufficient.

This chapter provides a concise overview of the main issues that should be considered in the design of representative mobile phone surveys. Constructing a sample is, in most ways, not much different in a mobile phone survey relative to other surveys, and many of the same resources are useful. Readers who would like to explore in greater depth the issues discussed in this chapter might consult, for example, *Survey Sampling* (Kish 1965), *Methodology of Longitudinal Surveys* (Lynn 2009), and *Practical Tools for Designing and Weighting Samples* (Valliant, Dever, and Kreuter 2013).

While this chapter highlights key issues, it is unlikely to provide sufficient depth on all relevant topics. The intricacies of individual designs are generally such that anyone preparing a mobile phone survey is advised to consult a sampling expert. This is a relatively small expense with, potentially, a high return because, in the end, inappropriately designed or weighted surveys are likely to generate misleading data and are, at best, of little value.

This chapter was written with Kristen Himelein.

Cross-Section or Panel Design?

The typical mobile phone survey will be a panel, that is, it will have a longitudinal design whereby the same respondents are interviewed multiple times. Cross-sections, whether repeated or stand-alone, are rarely an option because databases of mobile phone numbers from which a representative sample can be drawn are few.

The typical mobile phone survey, however, will start with a face-to-face baseline survey (or use an existing baseline survey), which is followed up with mobile phone interviews.[1] Once a set of respondents has been identified, it is cost-effective to continue interviewing the same respondents over time. Hence, a panel survey is created.

There are a number of benefits to using panel data. First, because the start-up costs are considerable for survey designs that begin with a face-to-face interview, continuing to interview the same respondents over time is cost-effective. Analytically, panel data are much more powerful than a repeated cross-section. Many mobile phone surveys are fielded to assess change because decision makers want to know the impact of a development program (for instance, who were the beneficiaries and losers because of a recent change in subsidies?), or because project leaders want to know the impact of their activities (for example, how many more people are sleeping under insecticide-treated mosquito nets compared with one year ago?). In some cases, researchers may want to establish causality (did opinions change after a major event?) or assess the cumulative effect of exposure (do fertility preferences among women change after the women watch soap operas for longer periods?). Panel data can be used to examine changes at the individual level between rounds, as opposed simply to the overall change in levels. Panel data are therefore more precise in measuring change than repeated cross-sections of the same sample size because one may control for respondent fixed effects. Thus, the elements of variation related to the region, village, household, individual, and so on remain constant between rounds.

The Implications of Frequency and Duration for Data Quality and Bias

In defining the frequency and duration of data collection, consideration of the trade-offs between analytical richness and potential sources of bias is important. Longer or more frequent panel surveys are able to capture more detail, but they are also more susceptible to biases and non-representativeness.

Depending on the rate of change in what is being measured, adapting the frequency at which information is collected may be necessary. More frequent data collection may be needed to understand the dynamics of a rapidly changing situation. If indicators change more slowly, less frequent data collection may be acceptable. The optimal frequency may even change over the course of a project. For example, in the early months of the Ebola outbreak, data were gathered more frequently because the situation was evolving rapidly. As the crisis shifted to recovery, regular data collection was still useful, but not at the same heightened frequency.

The frequency of data collection has quality implications, even though it is difficult to estimate the overall impact. Regular data collection can improve the quality of recall questions because the regularity provides a concrete bound on the recall period. Using a time marker, such as asking a question beginning with "since the last time we spoke..." can reduce error caused by telescoping or inclusion bias.

Surveys conducted at high frequencies may be accompanied by elevated nonresponse rates because of respondent fatigue (Wells and Horvitz 1978). In contrast, surveys that are carried out only rarely may be associated with higher nonresponse rates because respondents transfer phones, change phone subscriber identity module (SIM) cards, or lose interest in the process. The optimal frequency of data collection ultimately depends on the needs of the analysts, the behavior of the respondents, and the length of the overall data collection exercise. The typical MPPS survey collects information on a monthly basis, though weekly data collection is also possible. Even daily data collection may be feasible in certain circumstances.

Trade-offs also apply with respect to the overall duration of the data collection project. The longer the period covered by a panel survey, the richer and more valuable the data are likely to be. Longer duration panels, however, are susceptible to two main sources of bias. The first is related to attrition. As panel length increases, respondents are more likely to exit the survey. As long as a substantial number of respondents can be maintained and the survey still appears to be representative of the target population (which can be tested through nonresponse regressions, as discussed in chapter 5), there may be reason to continue the panel.[2] Even if the panel is no longer representative, there still may be reason to continue the project if the panel character is more important than the representativeness of the data, though this should be done with a full understanding of the possible impact on conclusions.

The second potential source of bias in long-term panels is panel decay. Panel samples are representative at the time they are selected, but begin to lose their representativeness almost immediately due to shifts in population. Taking the simple example of age, if there is no possibility of adding respondents to the survey, the average age of the panel respondents will rise irrespective of the average age in the overall population. To address this issue, some designs call for a panel to be updated by including "births" in the population since the original selection. These designs are called "fixed panels plus births." Another common design is a rotating panel, whereby a predetermined proportion of the sample units are replaced at regular intervals. Rotating designs are often used if the main objectives are short-term estimates of change and cross-sectional estimates over an extended period. These and other similar methods, including repeated panels and split panels, are less common in MPPSs than in other types of panel surveys, however, because the life span of an MPPS is typically no longer than two years.

In addition, there may be impacts on the behavior of respondents because of participation in a long panel survey. The reflective nature of humans makes it plausible that repeated questioning can change behavior (Sturgis, Allum, and

Brunton-Smith 2009). This can lead to biases that the literature refers to as "time in sample bias" or "panel conditioning." Panel conditioning needs to be taken seriously, particularly in MPPSs, because not only are respondents regularly asked to reflect upon their lives, but, in many cases, they also receive a free phone and call credit. This enhances the ability of respondents to communicate and to access information, and participation in the survey is therefore likely to have an impact on respondent's agency.[3]

Respondent Selection

An important early decision that should to be taken by the analyst is the unit of observation for the survey. While some mobile phone survey designs are not centered on households, but on, for instance, schools or firms as the unit of observation, most such surveys in the developing-country context are focused on households. In surveys that are designed to report only on household characteristics, the household is the unit of observation. In all rounds of the survey, as long as a respondent is able to speak about the household, it does not matter who responds. Respondents may even vary between the rounds of the survey, provided they are part of the same households. This practice is not recommended because it is likely to promote an increase in non-sampling errors if respondents have differing knowledge or perceptions about the circumstances of the same households. In addition, household itself may be a fluid concept if the number and composition of the members of the households change during the survey period.

Typically, all adults in a certain age range (15–65, for example) are eligible to participate in a mobile phone survey, but some surveys are limited to more specialized populations, such as university graduates, displaced persons, and so on. More often, mobile phone surveys collect information that pertains to individuals as well as households. If the survey is not targeting only household heads, the enumerator establishes a list of all household members during the baseline survey and determines who, among these, is eligible based on the selection criteria. If there are multiple eligible members, but only one respondent is needed, the enumerator randomly selects one household member from the list to be the target respondent. Whichever selection mechanism is used, the aim is to interview the same target respondents during every survey round. The selection of target respondents who will be available for the baseline survey and later phone rounds has the advantage of allowing for the participation of people with heterogeneity, that is, people with a wide range of characteristics. For example, the survey designers may want to learn about households, but also about migrant or mobile workers.

Sampling Frames

The sampling frame is the list of all possible units from which selected households are drawn. In an MPPS, even though respondents will be ultimately contacted by phone, the sampling frame might be based on phone numbers or on the household, or it might be drawn from an administrative list. The criteria that are important are that the sampling frame be exhaustive, not contain irrelevant elements, and not duplicate entries. This section discusses the possible options in more detail.

Phone Databases

In developed countries, households are selected to participate in a survey using Random digit dialing (RDD)[4] (for more information, see Massey, O'Connor, and Krótki 1997; Waksberg 1978). These approaches are effective in the developed-world context because nearly everyone has a phone number.[5] In developing-country contexts, mobile phone penetration is much less extensive, and access to personal landlines is often negligible. Moreover, the characteristics of individuals and households with phone access often differ greatly from those without such access. In particular, owners of mobile phones are more likely to be wealthier, live in urban areas, and be better educated (Pew Research Center 2015; Twaweza 2013; Wesolowski et al. 2012). Expenses related to charging phones and maintaining call credits exacerbate the gaps in the probability of survey response (Leo et al. 2015).

To draw representative samples through RDD, auxiliary information is required, and this might be another complication. A dataset that is known to be representative, such as a large-scale face-to-face national household survey or recent population census, could be used in combination with demographic and location information collected during the first round of an MPPS to create appropriate sampling weights. For example, if rural woman-headed households in a certain district were underrepresented in the RDD sample compared with the representative dataset, higher weights could be applied to these households to compensate. Generally, this requires some form of quota sampling whereby a minimum number of respondents is set for each category, and calls continue until this threshold is reached.

A more serious problem arises if characteristics correlated with low response rates are not readily measureable through an MPPS. Poorer households often have lower response rates, but poverty is conventionally measured with detailed income or consumption modules that are generally too long to administer as part of an MPPS. Imputation models based on proxy questions, such as assets or the occupation of the household head, can assist wherever direct estimation is not possible, but these methods usually introduce substantial error. This problem is compounded if certain categories of households are present only in insufficient numbers or absent entirely from random dial sampling frames. While aggregation with similar groups may be an option in certain cases, it is likely that the survey would not be considered representative for these groups.

For these reasons, RDD is not suitable for contexts in which the target population has low coverage rates and the aim is to create a representative sample.[6]

The Face-to-Face Baseline

As outlined in chapter 1, this handbook describes a design that begins with a (nationally) representative face-to-face baseline survey. This method is also used in the Survey of Income and Program Participation (see endnote 1).

The availability of a representative baseline removes much of the ambiguity related to the impact of nonresponse. Because detailed information, including wealth status, on the characteristics of those individuals or households that participate or do not participate is known from the survey, it is possible to use finer groups to calculate compensation weights for low response rates. The baseline survey can either be purposely collected for the project, as in the World Bank's Listening to Africa surveys, or rely on an existing survey that includes re-contact numbers.

If financially feasible, conducting a dedicated baseline has a number of benefits. First, using a previously collected baseline limits the total sample size and geographical allocation because only a certain number of households are available. Second, a dedicated baseline allows for the collection of specific variables relevant to the MPPS either as inputs to nonresponse calculations or directly in the analysis. Also, contact with future respondents provides the opportunity to explain the objectives of the survey, secure cooperation, and, if necessary, distribute free phones. This should raise response rates, particularly among the poorest. The baseline can also be piggybacked on a planned household survey. This could involve, for example, the distribution of phones and information to a subsample of respondents in the larger survey during the initial interview or, later, by a dedicated team. Regardless, if there are plans for an MPPS as a follow-up, current best practice dictates the inclusion of a re-contact information section in the questionnaire. This allows for flexibility and rapid mobilization during any unexpected subsequent crisis situation.

The methodology for conducting a face-to-face baseline for an MPPS is similar to the sampling strategy in any face-to-face survey. In most household surveys, no up-to-date sampling frame exists, and a frame must therefore be constructed. Typically, this is accomplished in two steps. First, clusters, usually census enumeration areas (EAs), are selected, generally through probability-proportional-to-size selection. For these areas, maps are obtained, and a listing exercise is carried out whereby basic information is collected on all households in the area. At the least, this needs to include information on the availability of mobile phone network coverage in the area, the size of each household, and other characteristics that the researchers need to collect to implement their sampling strategy. The information on mobile phone network coverage is subsequently used to decide whether the survey is viable in the area, while information about household size is used to determine appropriate sample weights. The list of households is used to draw the final sample of households and, eventually, to calculate analytical weights. (See chapter 3 for more detail on the preparation and implementation of baseline surveys.)

Other Sampling Frames

Some surveys are able to employ other lists of phone numbers as sampling frames. Completed household surveys can be a source of representative lists. This was the case in the World Bank's Liberia and Sierra Leone high-frequency phone survey during the Ebola crisis. Collaborating with institutions—ministries, unions, religious organizations, universities, or projects—that maintain databases with the phone numbers of the institutions under their authority or with which they collaborate, their members, their students, or their beneficiaries can also yield phone lists, though some consideration should be given to whether these lists have been updated and are representative of the population (box 2.1). In this case, as with RDD, the baseline phone survey must be carried out over the phone. This limits the number of questions that can be asked relative to a face-to-face interview, but, for many purposes, the information collected may suffice.

MPPS-Specific Challenges

Lack of Network Coverage

All mobile phone surveys assume that households are able to have access to mobile phone network. If some respondents live in areas without network coverage, and if the characteristics of these respondents are different from those of respondents in areas with network coverage, then a separate stratum may have to be defined in which respondents are interviewed face to face instead of through mobile phone interviews. While the inclusion of such a stratum will add to the complexity and cost of an MPPS survey, it is possible to limit the number of households included in the stratum to control costs. While all households must have a nonzero probability of being selected for a survey to be considered representative, the probability of being selected does not need to be the same provided sampling weights are used to correct the overall means. This under-sampling results in a loss in efficiency compared to an optimal allocation, but the results are unbiased. A survey that includes both mobile phone and face-to-face surveys also requires additional piloting to determine whether the change in survey mode affects the data.[7]

Box 2.1 Administrative Lists as Sampling Frames

Uwazi, the research unit of the Tanzanian nongovernmental organization Twaweza, has been conducting capitation grant surveys using administrative lists obtained from the Tanzania Teachers' Union (Uwazi 2010). The survey was aimed at assessing whether capitation grants had been disbursed to secondary schools. A sample of headmasters was selected from the teachers' union database. The results showed that the money had not arrived at schools, an issue that was quickly addressed once it became known. The impact of the monitoring exercise was significant, but the cost was negligible, less than $1,500.

Nonresponse and Attrition

Invariably, a share of the selected respondents will choose not to participate in the baseline survey (nonresponse) or will not participate in the first or subsequent cell phone rounds of a panel survey (attrition). If these respondents who choose not to answer are systematically different in any way, their nonresponse or attrition introduces bias into the estimates. While some of the bias because of nonresponse and attrition in the mobile phone survey may be offset (see below), complete correction of the bias is not possible. The only way to limit its impact is to prevent bias in the first place.

The optimal strategy for persuading respondents to participate in the initial round varies according to the context. General best practice involves maintaining a professional demeanor, addressing the questions and concerns of respondents, conveying the purpose and importance of the survey project, eliciting the backing of a reputable local organization and local leaders, and being respectful of the time commitment required of respondents. Sending reminder text messages prior to the survey phone call and telephoning initial non-respondents at a different time of day may raise participation rates. Many surveys also offer small incentives, such as mobile phone airtime credit, in appreciation of a completed questionnaire. (Financial incentives are discussed in greater detail in chapter 5.)

To maintain participation in subsequent rounds in a panel survey, the interview experience ought to be as pleasant as possible for the respondent. The same interviewer could conduct every interview, allowing the interviewer to build a rapport with the respondent. Dillon (2012) finds that matching a consistent interviewer with each respondent is useful for continued participation in the MPPS. A paired strategy may thus be more effective, whereby two interviewers rotate calling each respondent so that some rapport can be developed. This may also be a good way to check the quality of the data and ensure that interviewers do their job properly because they know another interviewer will do the interview during the next round and see the results from this round at least in the case of repeated questions.

Some panel respondents who dropped out of the surveys in Honduras and Peru cited the fact that they were asked the same questions frequently, as one reason why they dropped out of the survey (Ballivian et al. 2013). This is especially problematic for MPPSs with stable questions and relatively short intervals between survey rounds. This problem may be addressed by using dependent interviewing, which incorporates answers from earlier rounds of the survey in the current round. The Ebola impact survey in Liberia used the employment sector and job status of the respondent from the previous round in the current interview (Himelein 2014a). The question stated, "In the last round, you indicated that you were working in wage employment. Are you still engaged in this type of work?" Dependent interviewing reduces the respondent burden by eliminating redundant questions, thereby enhancing the flow of the interview. It also personalizes the interview and can help build rapport with the respondent. These types of questions additionally improve data quality both because a better relationship between the interviewer and respondent decreases nonresponse at the question level and because information is more likely to be consistent over time given that errors or nonresponse in previous rounds can be corrected.

For surveys taking place over longer periods of time, respondents can be kept informed of what has been done with the data collected to increase the level of engagement and goodwill. Many respondents appreciate learning through a text message or otherwise that the information they helped provide has been reported in a newspaper or on radio or television or that it was discussed in parliament. (For a more detailed discussion of various strategies to reduce nonresponse and attrition in an MPPS, see chapter 5.)

Replacing and Tracing

Regardless of efforts to retain respondents, some inevitably drop out. Statistical weighting methods can be used to minimize the impact of the resulting bias (see below). There is, however, a limit to how much correction can be accomplished using sampling weights. The replacement of respondents who no longer wish to participate is a common practice in many MPPSs, but this poses inherent risks to the validity of data. Unless the decision not to participate is completely random, replacement introduces bias. For example, if women are more likely to drop out of the survey, but they are replaced randomly from a 50 percent male–50 percent female population, the sample will become progressively skewed toward men. Some surveys attempt to match replacement households, such as replacing an older woman-headed household with a randomly selected household from a group with similar characteristics. This may lessen the impact of the bias, but the departing households are likely to be systematically different in unobservable characteristics as well. Moreover, this level of detailed replacement requires substantial information about the pool of potential replacements, which is rarely available. If the type of respondents who are more likely to drop out is known beforehand (the poorest, younger, and more elderly respondents tend to exhibit a larger propensity for nonresponse), the viability of the sample can be improved by oversampling among such types and then adjusting using sampling weights.

Tracing or tracking respondents, that is, using various ways to reach respondents by phone (for example, contacting the respondents through the references they provided during the baseline survey), can decrease attrition. In many ways, this is easier in mobile phone surveys than in traditional face-to-face data collection because the mobile phone travels with the respondent, while a face-to-face survey enumerator must physically locate respondents to conduct the interview.

This was a key reason an MPPS methodology was selected for the World Bank's Listening to Displaced People Survey in Mali (Etang, Hoogeveen, and Lendorfer 2015). Tracing in an MPPS context therefore centers on establishing protocols for those cases in which the originally selected target respondent is no longer in possession of the phone. It is best to plan for this situation in advance by collecting extra contact details during the baseline, especially phone numbers of household members, close friends, and, potentially, local officials who can be contacted if the target respondents become unreachable. Lists with such contact information need to be maintained and periodically updated throughout the process of data collection. An updated phone number for the target respondent may be available from the person who answers the call. This new number can then be added to the database. Analyzing the data associated with these respondents is more complicated than the analysis of data from non-mover respondents.[8]

If the person who answers the phone is a household member or close acquaintance of the target, and the target is still living in the household or only temporarily absent, it is also possible to ask that person to serve as a proxy and answer the survey questions in place of the target respondent. While this may decrease nonresponse, it can introduce other forms of bias (see Bardasi et al. 2011).

Another alternative is to replace the original target with a new respondent from the same household and proceed with subsequent rounds of the survey with the new respondent as the new target. This reduces the benefits of panel data because the analyst cannot consider the respondents as the same person when measuring changes, decreasing the overall sample size. This may also introduce bias if the new respondents are systematically different from the original randomly selected targets. Given these drawbacks, respondent switching should only be considered if maintaining the overall sample size is the most important consideration.

Depending on the goals of the survey, combinations of the above methods are sometimes used. For example, in the Liberia high-frequency survey, the target respondent was the household head. If the household head was not in possession of the phone, but the person answering was a household member, they could answer on behalf of the head. In situations where the respondent was completely new, most likely resulting because the initial interviewer incorrectly recorded the phone number, the person who answered became the new target respondent in subsequent rounds.

If multiple phone numbers have been collected for a given household, another common replacement strategy is, instead of retaining a mover respondent, to use the address or the household from which the respondent originates as the sampling unit. Provided not all household members have also moved, a replacement could then be found among any remaining, eligible household members. If all household members move, and the address is the sampling unit, an effort should be made to include the new tenants in the survey.

Sample Size Calculation

One of the most important issues that needs to be resolved in the design of an MPPS is the minimum necessary sample size. The relationship between sample size and cost is an important determining factor of whether undertaking the project is sensible. While sampling may become complicated as the design becomes more complex, and researchers should consult a sampling expert, a basic sense of the size of a sample can be obtained by answering the following seven questions.

1. What is the maximum acceptable margin of error? Typically, the answer is between 1 percent and 5 percent.
2. What confidence interval is sought? Typically, the answer is 95 percent.
3. What is the estimated (or prior) prevalence or standard deviation of the variable of interest? An example of the former might be that 30 percent of households should have at least one insecticide-treated mosquito net. An example of the latter might be the standard deviation of rice prices.
4. What is the size of the population from which the sample is being chosen? This becomes irrelevant for populations above 20,000.
5. What is the number of domains on which the survey is to report results? This has particularly important implications because the number serves as a multiplier on the minimum required sample size. For instance, if the survey only reports national-level results, a sample size of 500 may be sufficient. If the analysis plan calls for disaggregation by urban and rural areas, however, the calculations must be performed separately for each area, effectively doubling the required sample size. If a breakdown by gender within each of the rural and urban areas is required, then the number of domains increases to four, and the required sample size rises to 2,000. These issues can be particularly problematic in impact evaluations that have multiple treatment arms and interactions across these treatments.
6. What is the expected degree of attrition? If a panel design is being used, the final sample size needs to take account of the expected attrition during the life of the survey. This is important because even low levels of attrition can have significant cumulative effects on sample size. For instance, if the minimum required sample size is 500, and there are two domains (rural and urban), then the total pre-attrition sample size is 1,000. If the survey plan includes 24 rounds, if the attrition in each round is 1.5 percent, and if this attrition is random, then the eventual sample size would have to be 1,438 (calculated as $1,000/(1-0.015)^{24}$). Experience suggests that the greatest loss occurs between the baseline and the first round. If, between the baseline and the first round, 10 percent of the sample is assumed to be lost, and the target sample is 1,000 in round 24, then the eventual initial sample would have to be 1,573, calculated as $1,000/(0.9*((10.015)^{23}))$.
7. What is the impact of the panel design? Panel designs reduce the overall sample size requirements in the measurement of changes because, in panels, the same respondents are surveyed repeatedly, eliminating most systematic components of the variance. The degree to which the panel aspect of the data reduces the sample size depends on how much of the variance is related to the components, as measured by the correlation between the rounds of the survey. In practice, estimating this correlation at the sample design stage is difficult, and many therefore ignore this aspect of the calculation. Because it can only reduce the overall sample size required, ignoring it errs on the side of caution.

As a rule of thumb, one should expect that every domain requires 350–500 observations so that standard errors are about 5 percentage points around the mean or prevalence rate. These are percentage points rather than percentages. Five percentage points (or 10 percent standard error) for a true mean of 50 percent means that the confidence interval will be between 40 percent and 60 percent. The same 5 percentage points for a true mean of 10 percent means a confidence interval between 0 percent and 20 percent, and a 50 percent standard error. Estimating a 10 percent standard error for a 10 percent true mean would be 1 percentage point and require a much larger sample size.

Researchers who design their surveys to test whether an intervention leads to a significant result (that is, significantly different outcomes among those benefiting from the intervention and the control group) must perform power calculations. Power calculations estimate the minimum necessary sample size based on the expected difference in outcomes, the required level of confidence, and the variance in outcome variables.

A number of online calculation tools may be used to assist analysts in sample size calculations with the above parameters. Two widely applied tools are Optimal Design and Raosoft.[9]

Stratification and Clustering

In addition to the considerations above, sample size requirements are greater if the data are clustered. Stratification and clustering are common tools to gain greater control over the sample and to reduce costs.

Stratification involves dividing the population into mutually exclusive categories and treating each of the categories separately. Stratification has two objectives. First, it can be used to boost the efficiency of the sample through the inclusion of greater numbers of certain populations, which leads to greater heterogeneity. Oversampling in areas of high variation reduces the overall standard error. Second, stratification can be used to guarantee a minimum number of observations within a domain of analysis. A particular case is the level of reporting in sample size calculations described above. Each group on which independent estimates are calculated is considered a stratum.

In practice, these objectives can be contradictory, however. A country may have a large city with high variation, which, if it were to be oversampled, would reduce the overall standard error. The country may also contain a sparsely populated area with a high poverty rate, which would need to be oversampled to produce a sufficient number of observations to perform analysis at this level, but this would, at the same time, increase the overall standard error because of the low variation across the area. Stratification can thus either increase or decrease the standard errors.

Few face-to-face household surveys use simple random sample designs. In MPPSs, the sample selection and data collection costs are different relative to face-to-face surveys, and simple random sampling becomes a possibility. In designs that rely on RDD or an administrative database as the sampling frame and collect information only by phone, a simple random sample may be efficient. It would not require weights to compensate for differing probabilities of selection. A stratified simple random sample could provide even higher levels of efficiency, but weights would then be required. If the baseline is a face-to-face survey, however, households are likely to be clustered to manage survey costs. MPPSs have incentives to use clusters with relatively few households. The cost disadvantage inherent in a survey design with many small clusters applies only to the baseline (because the enumerators must travel more), but not to the subsequent rounds collected by phone. The additional travel costs during the baseline are a one-time expense, while the increase in precision resulting from the inclusion of additional clusters affects all rounds of the mobile phone survey.

Reducing the cluster size to 1, that is, no clusters or an unclustered design, may not be efficient in practice. Apart from the impact on cost, many mobile phone surveys use peer pressure to entice respondents to continue to participate. In the Sauti za Wananchi survey in Tanzania, for instance, all respondents in an EA select a group leader who, at times, is asked to follow up with respondents who cannot be reached (Twaweza 2013). Such possibilities do not exist or are more difficult if a completely unclustered design is used.

Clustering necessarily involves the selection of multiple households that are close to each other. This leads to greater homogeneity in the sample—people who live near each other have more in common than people who do not—and decreases the efficiency of the sample, while increasing the standard errors. Larger samples are needed to attain the same level of precision, but per interview costs are lower because enumerators travel less.

The final decision on the number of strata, the number of clusters, and the number of households within each cluster needs to be taken in consultation with a sampling expert. The optimal size of a cluster varies based on the objectives of the survey, but the general guideline is that smaller clusters are preferable for variables that are more highly correlated within clusters. For example, infrastructure surveys in where entire villages either have access to electricity or not have small cluster sizes. Surveys that cover socioeconomic issues, such as employment or poverty, have slightly larger clusters, between 6 and 10, generally. In surveys covering topics such as fertility or entrepreneurship that exhibit low correlations among neighbors, the clusters may be as large as 25. In stratification, the strata generally correspond with the domains of analysis, but more advanced sampling techniques increase the number of strata to reduce the standard errors (see Wolter 2007).

Sample Weights and Reweighting

Weight Calculation

Because MPPS samples are typically probability samples, weights are calculated upon the completion of the baseline to generate representative estimates. If the unit of observation is the household, that is, respondents are asked questions about the household as a whole, then the weights are at the level of the household. These weights allow for the unbiased estimation of the share of households meeting certain criteria or, if multiplied by the household size, the share of the population meeting the criteria. If the respondent is selected randomly from eligible household members and the questions pertain to that respondent as an individual, the weights incorporate the probability of selection and do not need to be multiplied by the household size.

Because data collection continues after the baseline survey and because different rounds show variations in the levels of nonresponse, round-specific weights need to be calculated for every interview round. Also, if the analyst intends to estimate a mean individual-level change between round 1 (wave 1: w_1) and round 2 (w_2) and if the analyst is forced to use a sample from the population consisting of units that existed at both w_1 and w_2 ($P_1 \cap P_2$), then appropriate weights must be calculated for this subset of households. These weights would only be appropriate for the comparison between round 1 and round 2. If the analyst wishes to repeat this exercise for the change between round 1 and round 3, separate weights must be calculated for the households appearing in these two rounds.

An important step in the (re)weighting procedures is the estimation of the likelihood of nonresponse using respondent characteristics. There is an extensive literature on respondent characteristics that are associated with nonresponse, and an MPPS project needs to ensure this information is collected during the baseline survey. This includes individual-level characteristics (gender, age, language ability, marital status, wealth) as well as information on the type of phone network used or the location of a respondent (table 2.1). A Stata procedure to calculate the appropriate set of weights from the base weights is presented in box 2.2.

Applying Weights

Survey weights are generally composed of the inverse probability of selection of the observation and a nonresponse adjustment (Himelein 2014b). In applying these weights, the weights must be appropriate to the statistic reported. For example, an MPPS that involves interviews among adults ages 18 and over could claim that 30 percent of the households do not have access to running water if the respondents spoke on behalf of the households. The same survey could find that 36 percent of the population does not have access to running water by using information on household size captured during the listing to transform household-level information into population information, effectively using population weights instead of household weights.

But the same survey, in reporting an opinion, might state that 20 percent of people ages 18+ do not to feel safe at home at night. This information cannot be presented as relevant for households or generalized to the overall population because the opinions of people ages below 18 have not been collected, and the sampling weights would include an additional term for the probability of selecting an eligible individual in a household.

In addition to the weights, surveys that have a complex design—those that incorporate stratification and clustering—need to apply this information in the calculations to estimate the standard errors correctly. In the first round of

Table 2.1 Baseline Survey Variables Typically Used for Reweighting Procedures

Variable	Survey response characteristics
Gender	Response rates tend to be higher among women than men
Age	Response rates tend to be lowest among the youngest and oldest in the sample
Language-speaking ability	Nonresponse is higher if the interview is carried out in a language that is not the mother tongue
Marital status	In cross-sectional surveys, single people exhibit a larger likelihood of nonresponse than married people; this may derive from the lower contact probabilities in face-to-face surveys
Household size and composition	Smaller households have lower contact probabilities in face-to-face interviews, and the evidence on refusal rates is mixed; people with children may be more likely to be home and more available for a face-to-face or phone interview
Education	Educational attainment is usually positively associated with survey response
Home ownership	Response rates are positively affected by this variable probably because it improves contact rates; the impact on mobile phone surveys is less clear
Income, wealth	Response rates are lowest in both tails of the distribution
Labor force status	The evidence on this variable is mixed, but, in some countries, labor force status has been found to be correlated with response
Location	Many surveys distinguish by region or by rural and urban location
Incentive	The value of the incentive appears to have limited impact on the response rate, though some incentives do show an impact
Phone network	Different phone networks vary in degree of reliability
Signal strength	The number of bars of signal strength during the baseline interview is correlated with response rates

Source: Adapted from Watson and Wooden 2009.

Box 2.2 Stata Code for Weight Calculations

The model starts with the variable *wt*, the household base weights, as well as a set of variables X that are expected to be correlated with nonresponse. The model first calculates nonresponse at the observation level, but then collapses the values into deciles. Appropriate weights (*wt_N*) for summary statistics for the data collected during round N can be calculated as follows:

```
logit response_dum_wave_N X
predict ps
xtile deca=ps, nq(10)
bys deca: egen ac=mean(ps)
replace ac=(1/ac)
gen wt_N=wt*ac                                        (B2.2.1)
```

The same procedure applies if the analyst uses a sample from the responses obtained in round N and round M [$P_N \cap P_M$], but response_dum_wave_N is now replaced by a dummy variable that takes the value 1 if responses are captured in both round N and round M, and zero otherwise.

---

the survey, this is straightforward because respondents are located in their original strata and clusters. As time passes, however, respondents may relocate to other strata and clusters. Fortunately for the analyst, the design information is frozen at the moment of selection so that respondents are considered part of their original cluster for the clustering of standard errors, even if they have physically relocated. A respondent who has moved from the countryside to the city would therefore be accompanied by the sampling information of the original location in the countryside, but report results based on their current location in the city.

Notes

1. MPPS design may be new in developing countries, but, in developed countries, such designs are common. An example is the U.S. Census Bureau's Survey of Income and Program Participation, a nationally representative longitudinal survey that collects data on income, wealth, and poverty in the United States. The first interview of the survey is administered in person, while subsequent interviews are generally conducted via telephone. Many business surveys are also set up as mobile phone panels.

2. However, these nonresponse analyses cannot rule out the possibility of systematic attrition on unobserved characteristics.

3. Twaweza, the Tanzanian nongovernmental organization that runs the Sauti za Wananchi (Voices of Citizens) survey, is in the process of investigating whether, after two years of participation in the survey, agency by respondents has changed relative to a control group. See "Sauti za Wananchi," Twaweza, Dar es Salaam, Tanzania, http://twaweza.org/go/sauti-za-wananchi-english/.

4. Random digital dialing refers to the use of all possible telephone numbers as a sampling frame for telephone surveys. Lepkowski (1988) provides a detailed review of RDD.

5. See Ferraro, Krenzke, and Montaquila (2008) on possible sources of bias in RDD surveys.

6. There are always exceptions. This was the case for the World Food Programme's mobile phone surveys on food security during the Ebola crisis in West Africa, where timing was more important than representativeness. Under these circumstances, analysts are encouraged to seek advice from an experienced statistician on the necessary modeling to mimic a representative sample.

7. See de Leeuw (2005) and Dillman (2009) for more information on mixed-mode surveys.

8. New information on household characteristics (location, housing, wealth, and so on) must be collected as covariates for the analysis and as inputs for any attrition corrections. The base probabilities of selection, excluding the attrition correction, remain constant from the time of selection, regardless of an individual's new circumstances. See Himelein (2014b) for details.

9. See, respectively, "Optimal Design with Empirical Information (OD+)," William T. Grant Foundation, New York, http://wtgrantfoundation.org/resource/optimal-design -with-empirical-information-od; and "Sample Size Calculator," Raosoft, Seattle, http:// www.raosoft.com/samplesize.html.

References

Ballivian, Amparo, Joâo Pedro Azevedo, William Durbin, Jesus Rios, Johanna Godoy, Christina Borisova, and Sabine Mireille Ntsama. 2013. "Listening to LAC: Using Mobile Phones for High Frequency Data Collection." Final report (February), World Bank, Washington, DC.

Bardasi, Elena, Kathleen Beegle, Andrew Dillon, and Pieter Serneels. 2011. "Do Labor Statistics Depend on How and to Whom the Questions Are Asked? Results from a Survey Experiment in Tanzania." *World Bank Economic Review* 25 (3): 418–47.

de Leeuw, Edith D. 2005. "To Mix or Not to Mix Data Collection Modes in Surveys." *Journal of Official Statistics* 21 (2): 233–55.

Dillman, Don A. 2009. "Some Consequences of Survey Mode Changes in Longitudinal Surveys." In *Methodology of Longitudinal Surveys*, edited by Peter Lynn, 127–40. Wiley Series in Survey Methodology. Chichester, West Sussex, United Kingdom: John Wiley & Sons.

Dillon, Brian. 2012. "Using Mobile Phones to Collect Panel Data in Developing Countries." *Journal of International Development* 24 (4): 518–27.

Etang, Alvin, Johannes Hoogeveen, and Julia Lendorfer. 2015. "Socioeconomic Impact of the Crisis in North Mali on Displaced People." Policy Research Working Paper 7253, World Bank, Washington, DC.

Ferraro, David, Tom Krenzke, and Jill Montaquila. 2008. "RDD Telephone Surveys: Reducing Bias and Increasing Operational Efficiency." JSM 2008: 1949–56. *Proceedings of the Section on Survey Research Methods, Joint Statistical Meetings*. Alexandria, VA: American Statistical Association.

Himelein, Kristen. 2014a. "The Socio-Economic Impacts of Ebola in Liberia: Results from a High Frequency Cell Phone Survey." World Bank, Washington, DC. https://hubs.worldbank.org/docs/ImageBank/Pages/DocProfile.aspx?nodeid=24048037.

———. 2014b. "Weight Calculations for Panel Surveys with Subsampling and Split-off Tracking." *Statistics and Public Policy* 1 (1): 40–45.

Kish, Leslie. 1965. *Survey Sampling*. New York: John Wiley & Sons.

Leo, Ben, Robert Morello, Jonathan Mellon, Tiago Peixoto, and Stephen Davenport. 2015. "Do Mobile Phone Surveys Work in Poor Countries?" CGD Working Paper 398 (April), Center for Global Development, Washington, DC.

Lepkowski, James M. 1988. "Telephone Sampling Methods in the United States." In *Telephone Survey Methodology*, edited by Robert M. Groves, Paul P. Biemer, Lars E. Lyberg, James T. Massey, William L. Nicholls II, and Joseph Waksberg, 73–98. Chichester, West Sussex, United Kingdom: John Wiley & Sons.

Lynn, Peter, ed. 2009. *Methodology of Longitudinal Surveys*. Wiley Series in Survey Methodology. Chichester, West Sussex, United Kingdom: John Wiley & Sons.

Massey, James T., Dan O'Connor, and Karol Krótki. 1997. "Response Rates in Random Digit Dialing (RDD) Telephone Surveys." JSM 1997: 707–12. *Proceedings of the Section on Survey Research Methods, Joint Statistical Meetings*. Alexandria, VA: American Statistical Association.

Pew Research Center. 2015. "Cell Phones in Africa: Communication Lifeline; Texting Most Common Activity, but Mobile Money Popular in Several Countries." Pew Research Center, Washington, DC.

Sturgis, Patrick, Nick Allum, and Ian Brunton-Smith. 2009. "Attitudes over Time: The Psychology of Panel Conditioning." In *Methodology of Longitudinal Surveys*, edited by Peter Lynn, 113–26. Wiley Series in Survey Methodology. Chichester, West Sussex, United Kingdom: John Wiley & Sons.

Twaweza. 2013. "Sauti za Wananchi: Collecting National Data Using Mobile Phones." Twaweza, Dar es Salaam, Tanzania.

Uwazi. 2010. "More Students, Less Money: Findings from the Secondary Education PETS." Policy Brief TZ.07/2010E, Twaweza, Dar es Salaam, Tanzania.

Valliant, Richard, Jill A. Dever, and Frauke Kreuter. 2013. *Practical Tools for Designing and Weighting Survey Samples*. Statistics for Social and Behavioral Sciences Series 51. New York: Springer Science+Business Media. http://link.springer.com/content/pdf/10.1007/978-1-4614-6449-5.pdf.

Waksberg, Joseph. 1978. "Sampling Methods for Random Digit Dialing." *Journal of the American Statistical Association* 73 (361): 40–46.

Watson, Nicole, and Mark Wooden. 2009. "Identifying Factors Affecting Longitudinal Survey Response." In *Methodology of Longitudinal Surveys*, edited by Peter Lynn, 157–82. Wiley Series in Survey Methodology. Chichester, West Sussex, United Kingdom: John Wiley & Sons.

Wells, H. Bradley, and Daniel G. Horvitz. 1978. "The State of the Art in Dual Systems for Measuring Population Change." In *Developments in Dual System Estimation of Population Size and Growth*, edited by Karol J. Krótki, 53–73. Edmonton, AB, Canada: University of Alberta Press.

Wesolowski, Amy, Nathan Eagle, Abdisalan M. Noor, Robert W. Snow, and Caroline O. Buckee. 2012. "Heterogeneous Mobile Phone Ownership and Usage Patterns in Kenya." *PLoS ONE* 7 (4). http://journals.plos.org/plosone/article?id=10.1371/journal.pone.0035319.

Wolter, Kirk M. 2007. "The Method of Random Groups." *In Introduction to Variance Estimation*, 21–106. Statistics for Social and Behavioral Sciences Series. New York: Springer Science+Business Media.

CHAPTER 3

The Baseline Survey

Introduction

This chapter presents key considerations to help guide the implementation of the baseline for a mobile phone panel survey (MPPS). A first aspect to resolve is whether the survey will be implemented by an internal team or is to be outsourced. This chapter is relevant no matter which option is selected. Those who implement each step of an MPPS internally can use the information as a guide, and those who are interested in outsourcing the baseline survey may use the information in monitoring the quality of the baseline survey.

Groundwork: The Distribution of Mobile Phones and Solar Chargers

The baseline survey is the foundation upon which an MPPS is built. The baseline survey must therefore be conducted with the utmost care and attention. The success or failure of a mobile phone survey is heavily dependent on the successful implementation of the baseline survey.

Although mobile phone ownership is growing rapidly, many countries remain far from achieving universal mobile phone ownership.[1] For example, in the Tanzanian case, evidence in the baseline surveys Listening to Dar and Sauti za Wananchi show that mobile phone ownership is more widely distributed among richer households than among poorer households (figure 3.1). This is consistent with findings from almost all national household surveys across many countries.

As part of the groundwork in contexts in which access to mobile phones is not universal and ownership is associated with socioeconomic status, it is good practice to distribute mobile phones to respondents to ensure the representativeness of the survey results. The distribution might occur according to one of the following strategies:

1. *Provide mobile phones to all respondents*: In favor of this method is the argument that receiving mobile phones and chargers incentivizes participation in the panel. Moreover, distribution of mobile phones to all respondents whether

Figure 3.1 Household Mobile Phone Ownership, by Wealth Quintile, 2010–12

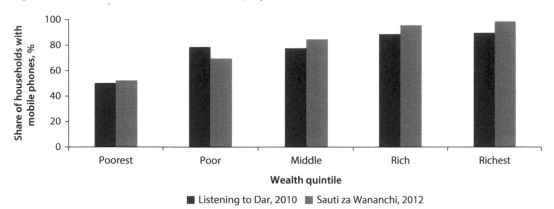

Sources: Baseline surveys: "Listening to Dar," Twaweza, Dar es Salaam, Tanzania (accessed August 22, 2014), http://www.twaweza.org/go /listening-to-dar; "Sauti za Wananchi," Twaweza, Dar es Salaam, Tanzania (accessed August 22, 2014), http://twaweza.org/go/sauti-za-wananchi -english/.

these already have mobile phones or not ensures the representativeness of the sample across all households irrespective of wealth status.

2. *Provide mobile phones only to those respondents who do not already own mobile phones*: This decision is mainly driven by narrow budgets. If opting for this method, one should be aware that it may lead to demoralization among the MPPS respondents who do not receive new phones. To avoid potential conflict, in enumeration areas (EAs) in which no MPPS respondents own a mobile phone, one might consider distributing mobile phones to all respondents or giving them something of similar value.

The distribution of solar chargers is also an important consideration, especially in the case of respondents who live in households that do not have a reliable supply of electricity. This is important because, without access to the electricity grid, one of the main reasons respondents cannot be reached by interviewers is that the phones of the respondents have not been recharged (see chapter 5, figure 5.4).

Considerations in Purchasing Mobile Phones and Solar Chargers

Should it be decided to distribute mobile phones and solar chargers, the most suitable hardware must be selected, as follows:

1. *Ease of use*: Especially in a developing country, a representative MPPS is likely to cover people from diverse backgrounds in terms of educational attainment, income levels, and other characteristics. The mobile phone handsets and solar chargers chosen must therefore be technologically accessible to all types of people, but particularly people with little or no education or experience

with technology. The visit of survey enumerators at each household is short. A user-friendly device would allow the enumerators to teach the respondent about the functions of the mobile phone quickly before moving on to the next household.

2. *Value for money*: A major motivation for the implementation of MPPSs for high-frequency data collection is the cost-effectiveness of the methodology, which cuts out the cost of repeated travel to the field to conduct face-to-face interviews. To take advantage of the cost savings of the mobile phone handsets and the solar chargers, the implementing partners might seek a balance between the quality of the device and the cost. They should try to purchase hardware covered by a warranty. This ensures that attrition does not occur because of hardware failure not covered by warranty.

3. *Durability*—An MPPS is a longitudinal study that may last two to four years. Implementers must procure durable devices that last beyond the project life cycle.

4. *Battery life*: Both the solar charger and the mobile phone should have batteries with appropriate capacity and durability. The solar charger battery should be able to store a reasonable amount of power, and the phone battery should be sufficient to maintain phone operation and ensure the availability of respondents for the calls of interviewers.

5. *Simple phone*: It may be unwise to distribute attractive handsets to household members who lack status within the household because they may be less likely to keep such handsets.

In selecting a mobile phone handset, one might also consider the following:

1. *The appearance and screen color*: MPPSs include respondents in various social classes and in rural and urban areas. The handsets might be selected to appeal to all groups or respondents might be offered a choice of colors and brands of comparable quality to ensure that no one feels deprived or disadvantaged. One might also consider giving other gifts of similar value to respondents who already own mobile phones.

2. *Tracking features*: To recover stolen phones, but also to minimize the risk that phones are sold on by survey participants, the implementing partners might consider acquiring phones that have a tracking feature. If this is done, respondents should be informed of the tracking feature and the purpose of the feature. The cost of actually tracking a phone, which may require police involvement, may outweigh the benefits.

Handling the Registration of SIM Cards

Communication regulators across the world require the registration of mobile phone numbers. Many countries, including most of Sub-Saharan Africa and a number of countries in Latin America, now require users to register their subscriber identity module (SIM) cards (GSMA 2013).

Mobile Phone Panel Surveys in Developing Countries • http://dx.doi.org/10.1596/978-1-4648-0904-0

In the event that mobile phone handsets are distributed to new mobile phone users, then the phones should come with functioning SIM cards. All SIM cards must be registered either during or immediately after the baseline survey by the survey implementer or researcher. The SIM cards may be registered using the names of the respondents, if they so prefer. Some respondents who already have SIM cards may not want an additional one.

To ensure the success of the project, the survey implementers might partner with the local mobile network operator in countries in which there is a single competent operator (box 3.1). In more competitive markets or countries, implementers might consider partnering with several network operators. This may help in addressing possible SIM card rejections as well as network issues given that some operators may have better network connections and usage than others in some EAs.

Choosing the Mode of Data Collection during the Baseline Survey

Various modes of data collection are available for conducting effective face-to-face interviews. These include paper-and-pencil interviewing (PAPI) and computer-assisted personal interviewing (CAPI).[2]

In MPPSs, the selection of the data collection method might best be aimed at reducing the lag time between the baseline survey and the first mobile phone survey and maintaining about the same lag times between successive mobile phone survey rounds. Experience has shown that the longer the lag time between successive phone survey rounds, the lower the response rate in each of the successive rounds. For example, in Madagascar, the response rate during the first four phone survey rounds in 2014 ranged between 93 percent and 96 percent (figure 3.2). It dropped to 76 percent in round 5 in 2014 and remained about the same up to round 12 in September 2015. The widening lag arose because of a change in financial arrangements in the midst of the project. By the time new funds had been secured to continue the survey, a substantial number of respondents had withdrawn or were no longer reachable. The project team adopted various strategies to try to fix the problem, including field revisits to the more

---

Box 3.1 Tanzania Mobile Phone Survey

Sauti za Wananchi, the Tanzania mobile phone survey, partnered with three of the leading mobile network operators in Tanzania, Airtel, Tigo, and Vodacom. Given that the survey team did not know who would be randomly selected as respondents in the panel survey, the enumerators were trained in SIM registration and carried registration forms along on their visits to households for the baseline interviews. They filled out these forms during the interviews and returned them to the appropriate network operator.

Source: "Sauti za Wananchi," Twaweza, Dar es Salaam, Tanzania (accessed August 22, 2014), http://twaweza.org/go/sauti -za-wananchi-english/.

Figure 3.2 Response Rates, Listening to Madagascar, Rounds 1–18, 2014–16

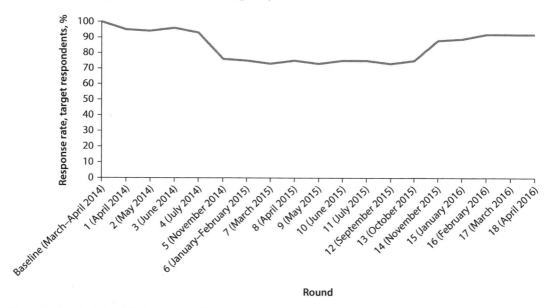

Source: Data from the fieldwork for the Listening to Madagascar survey.

difficult to reach households that were conducted in October–November 2015. This approach clearly paid off. The response rate started to increase immediately after the field revisits and stood at 92 percent in April 2016, two years after the start of the mobile phone survey.

Ideally, the first round of the mobile phone survey should be run alongside the baseline survey so as to eliminate or reduce drastically the lag time by immediately engaging with the respondents. To create an overlap between the baseline survey and the mobile phone survey rounds it is good practice to share identification information on the respondents with the call center during the baseline data collection period. The most efficient way to accomplish this is to rely on technology to transmit real-time field data to the call center or other central location by text or otherwise, or, if the data collection is conducted relatively close to the headquarters of the survey implementers, the paper questionnaires could be taken there.

If the PAPI option is selected for data collection, then, to overlap the baseline survey and the first phone interviews, the field enumerators ought to have a means of sending information on a few respondent variables to the central hub, for example, by short message service (SMS). These key variables might include the following:

- Name of the respondent
- Telephone number
- Household identifier
- Unique EA identifier
- Questionnaire field serial number

The call center team at the hub can create an interim database using the key variable information and immediately begin telephoning respondents, thereby bridging the lag between the baseline and the first mobile phone survey call round.

As an alternative to running the baseline survey and the first round of calls simultaneously, the field team could make brief courtesy visits to respondents a week or so after their first contact, that is, after the baseline survey. This would mimic the courtesy social visits that are customary in many countries and would also be an opportunity to tie closer attachments to the respondents.

Obtaining Official Licensing for the Survey

Each country has policies to regulate research activities, including research through surveys. If it is required, the survey implementers should obtain official approval, licensing, and certification for the survey in a timely manner before the planned launch of fieldwork. The licensing process may take several months because of the hurdles inherent in government bureaucracy. Depending on the purpose of the survey and local regulations, permits may be needed from more than one research clearance authority. For example, in Tanzania, clearance must be sought at the Commission on Science and Technology, the Ministry of Education, and the Ministry of Health before studies can be conducted in health care facilities and schools.

Before surveys can be conducted, additional permits may also be required as part of a country's human subjects research ethical procedures. These might involve, for instance, approval from a governmental institutional review board or, if local university researchers are participating, clearance from the university review board. The implementer and researchers are responsible for ensuring compliance with all regulations and ethical standards associated with relevant local review board procedures.

MPPS Instruments: The Baseline Survey

The MPPS baseline survey involves many steps, each requiring special preparation. Thus, data collection tools must be readied; enumerators and interviewers trained; respondents identified, contacted, and coached; communities informed and engaged; and fieldwork carried out and monitored. The various survey instruments that are particular to an MPPS are described below, including functions and key issues. The list is focused on the needs of a household survey and is not exhaustive. Additional or fewer instruments may be required, depending on the purpose, nature, and context of the specific survey.

Household Listing Form

The implementers must establish lists of all households in each EA of the mobile phone survey if such lists are not otherwise available. The listing exercise is crucial because it provides all households in every EA with an equal chance during the random selection of households for the survey. A listing form is used in this listing exercise. The listing form is designed to capture the name of the household head, the household size, the number of adult household members, mobile phone ownership among household members, network accessibility at each household, and all other information on each household in the EAs.

Head of Household Consent Form

In an MPPS that requires the participation of household members other than the heads of household, the research team should seek the consent of the household heads for the phone interviews among other members (box 3.2). The consent form ought to describe the purpose of the survey, the procedure, the random selection of adult respondents, policies on data confidentiality, and benefits and risks, if any. The form is filled out and signed in duplicate. One form remains with the respondent, and the second form is retained by the implementers. Appendix A offers an example of the head of household consent form.

Respondent Agreement Form

It is necessary to keep a copy of the head of household consent form, as well as the respondent agreement form. The agreement form documents the agreement between the MPPS implementer and the respondent. The agreement form sets out the rules of the interactions between the two parties during the research. The implementers might consider including the following information in the form: an overview of the survey, a respondent confidentiality clause, the panel timeline, the purpose and conditions of the distribution of mobile phones and solar chargers, the ownership of the mobile phone and solar chargers, and any incentives or benefits the respondent might expect arising from survey participation. As with the head of household consent form, this form should be filled out and signed in duplicate. Unlike the head of household consent form, it is necessary to have physical documentation of the respondent agreement. Appendix B contains an example of a respondent agreement form.

Box 3.2 Intrahousehold Conflict and the Head of Household Consent Form

The head of household consent form was adopted in the early stages of rolling out the Tanzania nationally representative survey, Sauti za Wananchi. During the first week of the baseline survey, the team learned of reports about two cases of women beaten by the head of household husbands for accepting mobile phones from strangers. To prevent other such incidents, the research team decided to seek consent from heads of household before they randomly selected household members for interviews. This resolved the problem.

Source: Twaweza 2013.

Respondent Training Manual

The mobile phones and solar chargers may be distributed to first-time users. Developing a respondent training manual on how to use these devices may therefore be advisable. The training manual might incorporate lessons learned during the pretest of the baseline questionnaire. It might also serve as a didactic tool for enumerators as they engage with respondents during the preparations for the baseline interview.

The purpose of each MPPS may differ from one to the next; so, training manuals ought to be customized. For example, if the implementers want respondents to undertake specific activities and report the results over the phone, such as measurements of the weight and height of children, then this needs to be spelled out in the manual.

Communication Materials

In most settings, respondents are not familiar with the concept of an MPPS and the distribution of mobile phones and solar chargers for research purposes. Information gaps therefore exist that, if not addressed carefully and in a timely fashion, might lead to misunderstandings and unease in the community or high attrition rates among respondents. For this reason, materials might be developed to lay out this sort of relevant information to share with community members in the selected EAs. Such an information campaign would target the entire community, that is, it ought not to be limited only to the target respondents. For example, in the Tanzania MPPS, the team distributed booklets containing a story presented in cartoons that explained the key details of the panel survey (Kamwaga and Mushi 2012). In Madagascar and Zanzibar, a short video was shown to respondents during the baseline survey. In the Madagascar video, the director general of the National Statistical Office explained the objectives and importance of the mobile phone survey and encouraged the selected households to participate. This information campaign was quite useful in promoting survey participation.

Checklist

An MPPS involves more effort than a traditional survey. This is mainly because of the number of extra steps that the implementing research team must take to ensure that they build a mobile phone survey that can last through several rounds. For this reason, the implementing research team might consider developing a checklist to verify that all necessary preparations have been completed, for example, Are all the survey tools ready? Have enumerators been trained? Has the community information campaign been rolled out? The checklist should be developed and discussed among the core implementer team. It can be a tool for monitoring progress and assigning responsibilities. Appendix C provides a checklist of some key items, stages, and activities the completion of which may be essential before the rollout of the baseline survey is possible. The list is far from exhaustive, though it does represent an initial template.

Piloting an MPPS

A pilot survey is a mirror of a full survey. An MPPS ought to be piloted before it is implemented. The inputs of the pilot survey allow the research team time to discover gaps, fine-tune the survey instruments, and tweak the survey interview scheme before interviewing the target sample. The enumerators who participate in the pilot survey might consist of a mix of experienced and less experienced enumerators, thereby allowing the pilot survey to become also a training exercise. Possible steps in piloting an MPPS include the following:

1. Establish the number of days needed to complete the baseline survey exercise in a single EA.
2. Establish the practicability of connecting respondents with the call center.
3. Test the household listing process and the random selection of households and respondents.
4. Test the planned strategies to manage attrition, specifically:
   a. The distribution of mobile phones and SIM cards as a tool for data collection
   b. The registration of new SIM cards
   c. The use of solar chargers to facilitate the charging of phones
   d. The organization of respondents into groups
   e. The use of group leaders and group partners within groups
5. Assess the attrition rates among pilot respondents in a preliminary round.
6. Identify any challenges in fielding the pilot baseline survey that need to be addressed through enumerator and interviewer training or during the implementation of the full baseline survey.
7. Evaluate the functioning of any revised survey instruments.
8. Estimate the number of EAs that may experience mobile phone network issues.
9. Test the CAPI application.

Pilot EAs should be randomly selected in both rural and urban areas, thereby offering an appropriate balance for the identification of possible challenges in each environment. Pilot EAs should not form part of the main survey sample, though they may receive the same treatment as the target sample, including the random selection of respondents and mobile phone distribution. Given the potential need to test the survey instruments in any future round of the survey or in any future survey using the same EAs and household lists, the pilot EAs, if they are maintained as a small survey subsample distinct from the target sample, may be reused to pretest new or revised MPPS questionnaires.

Building Team Spirit and Community Engagement

The success of an MPPS depends on the ability to reach as many of the baseline target respondents as possible as often as necessary. Organizing a group meeting with all randomly selected respondents in an EA may therefore be advisable before the research team leaves the EA. The list of aims that such a group meeting might serve might include the following:

- Provide resources for and training in the use of solar chargers. This may also help incentivize attendance at the meeting.
- Address concerns and questions about the MPPS.
- Reassure respondents that all survey responses are correct answers and that respondents neither benefit nor are punished for their responses.
- Select a group facilitator or group leader to help trace or track respondents and to collect community monitoring data on behalf of the research team, for example, whether the public water utility continues to supply improved water sources.
- Identify and pair neighboring respondents so as to facilitate the tracing of respondents through neighbors.

Groups of respondents organized around appointed leaders might be more useful in rural areas than in urban areas because social control and the authority of community leaders are often less pronounced in urban settings.

The concepts behind an MPPS and the distribution of mobile phones are probably unfamiliar in most target communities. Communication materials distributed among the respondent groups can help explain the purpose of the project (see above). These materials might include booklets and short videos. Group or community sensitization meetings as part of an awareness campaign might also be effective. The 2014 panel of the Sauti za Wananchi MPPS in Tanzania demonstrated that community meetings are crucial in helping build trust, demystifying any impression of bias, and enhancing survey buy-in.[3] Depending on the MPPS budget, the research team might also consider engaging local media outlets in the effort to inform target communities.

Challenges Associated with the MPPS Baseline Survey

Researchers conducting field surveys face numerous challenges, ranging from transport issues to the development of the survey questionnaire and the interpretation and cataloguing of responses. This section focuses on the fieldwork challenges specific to an MPPS.

Suspicion of the Aims of the Field Team

Communities may distrust the intentions of the survey implementers and researchers. Even the distribution of free mobile phones and solar chargers may arouse suspicion (Twaweza 2013). For example, in Kenya, Madagascar, Malawi, and Tanzania, communities that linked the gifts with attempts by fraternal organizations to win their support resisted the implementation of the surveys. A few target respondents dropped out of the surveys immediately after joining because of these concerns.

To prevent the spread of this and similar false rumors by community members who are not sampled and therefore may feel excluded, research teams explained the process of random sampling to local group leaders and during community meetings in Tanzania in 2015. Teams might be proactive and engage on such issues with groups and communities more fully by providing information booklets and other communication materials that clarify the aims of the surveys (see above). They might also put stickers on the phones that define the purpose and source of the phones, thereby weakening false associations and strengthening the direct association between the free phones and the research goals. Subsequently, after the launch of a survey, the teams might offer evidence of the data outputs.

Household Conflict over Mobile Phone Ownership
A mobile phone is a valued asset in poor households. The act of handing out free mobile phones to women household members or younger household members while the heads of household do not have a mobile phone can lead to conflict over ownership and control (see above). For this reason, enumerators must seek the consent of the heads of household before selecting target respondents from the households. Failure to do this could put the well-being of respondents at risk.

Network Access
Network reception is crucial to the success of a mobile phone survey. Research teams typically seek to exclude EAs in which there are mobile network issues or revert to field-based surveying to collect the data in these areas. Border areas are likely to exhibit network overlaps between neighboring countries, which might foster roaming, particularly if the network operators in a neighboring country possess stronger coverage. Target respondents might avoid answering their phones to avoid exorbitant fees under such conditions. The research teams might consider providing respondents in these areas with incentives to offset these potential fees.

Conclusions
The above list of challenges is far from exhaustive. The objective is to highlight a few challenges, particularly those that have been encountered in the field. In considering possible barriers, implementers and research teams need to tailor the MPPS data collection system to the local context (Ganesan, Prashant, and Jhunjhunwala 2012).

One way to minimize the risk of nonresponse is to collect alternative phone numbers of the target respondents during the baseline. These may help trace the respondents or their households. Beware, however, that the collection of alternative contact numbers does not have implications in terms of institutional review board regulations. Prescheduling survey contact days and times during the baseline may raise response rates. Network access may also sometimes be partial or intermittent; thus, for example, the footprint of a network might shrink during peak traffic periods. Prescheduling or sending reminder SMSs out before the interviews can alert willing respondents to go to locations where there are better signal strengths. Indeed, an SMS is complementary to an MPPS because it facilitates brief or emergency contacts sufficient for making scheduling changes or transmitting reminders to respondents. However, there may be extra costs associated with use of an SMS, though, in most places, these are likely to be minimal relative to the costs of nonresponse or attrition.

Notes

1. See Ballivian et al. (2013). See also "Mobile Cellular Subscriptions (per 100 People)," World Bank, Washington, DC (accessed August 21, 2014), http://data.worldbank.org /indicator/IT.CEL.SETS.P2.

2. PAPI is a traditional method whereby data are collected in the field on paper questionnaires and entered into the database either using a data entry application or scanning (rare) and either at headquarters or a regional office. CAPI refers to the method of collecting data on tablets or personal digital assistants using interactive software and regularly transmitting the data to headquarters using the mobile phone network.

3. See "Sauti za Wananchi," Twaweza, Dar es Salaam, Tanzania (accessed August 22, 2014), http://twaweza.org/go/sauti-za-wananchi-english/.

References

Ballivian, Amparo, João Pedro Azevedo, William Durbin, Jesus Rios, Johanna Godoy, Christina Borisova, and Sabine Mireille Ntsama. 2013. "Listening to LAC: Using Mobile Phones for High Frequency Data Collection." Final report (February), World Bank, Washington, DC.

Ganesan, Muthiah, Suma Prashant, and Ashok Jhunjhunwala. 2012. "A Review on Challenges in Implementing Mobile Phone Based Data Collection in Developing Countries." *Journal of Health Informatics in Developing Countries* 6 (1): 366–74.

GSMA. 2013. "The Mandatory Registration of Prepaid SIM Card Users: A White Paper, November 2013." GSMA, London. http://www.gsma.com/publicpolicy/wp-content /uploads/2013/11/GSMA_White-Paper_Mandatory-Registration-of-Prepaid-SIM -Users_32pgWEBv3.pdf.

Kamwaga, Evarist, and Elvis Mushi. 2012. "Je, nini kipya? Sauti za Wananchi: njia poa ya kusikika!" [What's new?: Sauti za Wananchi, a cool way to be heard!] With illustrations by Marco Tibasima. Twaweza, Dar es Salaam, Tanzania. http://www.twaweza .org/uploads/files/Sauti%20za%20wananchi(1).pdf [in Swahili].

Twaweza. 2013. "Sauti za Wananchi: Collecting National Data Using Mobile Phones." Twaweza, Dar es Salaam, Tanzania.

CHAPTER 4

Setting Up a Call Center

Introduction

This chapter describes the decisions and steps to be taken in setting up a call center for a mobile phone panel survey (MPPS). Although this handbook discusses the baseline survey and the mobile phone survey rounds in separate chapters, MPPS projects in Latin America, South Sudan, and Tanzania have demonstrated that the call center and the baseline interviews of an MPPS project ought to be planned and implemented in parallel rather than consecutively (Ballivian et al. 2013; Demombynes, Gubbins, and Romeo 2013; Hoogeveen et al. 2014). This parallel implementation allows the first mobile phone contact to be established between a target respondent and the call center while the baseline enumerator is still at the respondent's household. The early contact tends to reinforce the association for the respondent between the call center and the first round of the mobile phone surveys. Experience shows that MPPS projects that manage a seamless transition between the baseline interview and the phone interviews are more successful in reducing the significant attrition that typically occurs after the baseline survey and before the first round. Indeed, projects with a delay between the field visit and the first call typically struggle with much higher attrition rates throughout (see Hoogeveen et al. 2014). The contact also allows the baseline enumerator to help the respondent with any technical problems that might otherwise occur during the first phone interview, especially if the respondent is unfamiliar with the use of mobile phones.

Building an In-House Call Center vs. Outsourcing

A fundamental decision in planning the mobile phone survey phase of an MPPS project revolves around the choice between setting up a call center in-house or hiring a professional call center to conduct the interviews.[1] The decision whether to build or outsource depends, first, on the skills and resources available to the project. The time and cost implications of the two options usually vary considerably (table 4.1). In particular, building a call center from scratch can be time-consuming, and the costs are often less predictable. Meanwhile, though the

Table 4.1 Pros and Cons of Building or Outsourcing a Call Center

Pro or con	In-house call center	Outsourced call center
Advantage	• More control over the entire research process, including staff recruitment and training • Easier to safeguard data confidentiality and data security • In-house capacity building	• Ability to make use of experienced project managers and call center interviewers • Professional hardware and software are readily available, allowing for more sophisticated quality control and protocols to reduce nonresponse • Allows the project leader to focus on data analysis and dissemination rather than on data collection
Disadvantage	• Investments may be unjustified for a single MPPS project • Time-consuming recruitment, training, and management of the call center team • Often requires the involvement of external consultants, for example, the installation of software and hardware, staff training, and data management	• Typically, more costly • Less flexibility in making short-notice changes to work flows and questionnaires

start-up costs of the ad hoc call center may appear high, the fixed operational costs associated with a professional call center are typically higher in the long run. Presumably, professional call centers are more costly because they need to make a profit, and their greater efficiency and timely service delivery relative to a call center established in-house or at a national statistical office come at a cost.

Nonetheless, setting up a call center in-house and training the staff of interviewers can represent a valuable investment in an organization's future capacity to run MPPS projects and respond quickly to emerging data needs. In addition, because of the nature of the data to be gathered, the social and political environment, and local privacy and data protection regulations, the implementing organization might opt for full control over the entire flow of data gathering, management, analysis, and dissemination.

In the following section, guidelines for setting up an in-house call center are described. While most of the remainder of the chapter thus assumes a situation in which some type of in-house call center is established, the chapter may also still provide some useful general quality control measures for projects in which the phone interviews are outsourced.

Technology Considerations

In creating the infrastructure of an in-house call center, an implementer is faced with a wide range of hardware and software options. At the low end of complexity and sophistication, a call center consists of office space equipped with mobile or fixed phones and devices with a simple data entry mask, such as Microsoft Access, Microsoft Excel, or IBM SPSS Statistics. Devices for data entry might range from desktop personal computers and laptops to tablets, personal digital assistants, and even mobile phones. In this simpler scenario, interviewers manually dial a respondent's phone number, record responses during the interview using the data entry interface, save the data, and move on to the next phone call.

At the most sophisticated end of the scale, call center work flows are largely automated. Computer-assisted telephone interviewing (CATI) software is used to integrate the interview process and data entry. This leaves less room for data entry mistakes and provides call center supervisors with an arsenal of quality control measures. In this scenario, a centralized call center software automatically assigns respondents to interviewers, based on the time preferences of the respondents and, possibly, other matching criteria such as language.[2] The software loads a phone number, calls the number after prompting by the interviewer, presents the interviewer with the questionnaire (typically, one question per screen), and provides data entry fields under each question.

In this scenario, data entry and the phone calling process are converged into one device such as a desktop computer, and interviewers wear headsets, which allow them to record data more accurately and quickly. When the interview commences, the interviewer has access to relevant information, such as the respondent's sociodemographic background characteristics or past nonresponse behavior. This means that interviewers can use variables such as age, name, and gender to confirm they are talking to the target respondent, while editing sociodemographic background information if changes have occurred, such as birth of a child. Having quick access to the past responses of the respondent also allows the interviewer to ask more tailored and targeted questions. For instance, a question about the school attendance among children in the household can be asked in a more personal manner, with information available about the number and names of the children.[3] Supervisors can monitor the progress of individual interviewers, listen in to live or recorded interviews, and produce preliminary data reports, while a data gathering round is ongoing.

An infrastructure such as the one described in the second scenario is typically found in professional call centers or larger-scale MPPS projects. However, an immense variety of open-source and proprietary call center and data entry software packages are now available, and even MPPS projects without the resources to build a full-fledged professional call center infrastructure can benefit greatly from such software in managing work flows and quality control. The correct choice of a software solution is crucial in any MPPS project.

Relying on a software solution with centralized data management, that is, all data entry personnel feeding into a single database, will almost always be the preferred option. This allows supervisors to extract data in real time and to track the progress of individual interviewers to see if response-rate targets are being met.

Most modern data entry software also allows for routing, that is, automatically skipping certain questions conditional on the responses to earlier questions. This permits more complex questionnaire designs and reduces the risk of mistakes that may arise if question routing is manual. In addition, some software packages can be used to conduct data consistency checks and spot obvious mistakes, for instance, that a 7-year-old is recorded to have a child.

An issue closely related to the software decision is the choice of data entry hardware. Some software interfaces are more suitable for personal computer screens than for tablet touchscreens, and vice versa. Other software packages can be used cross-platform.

The Recruitment and Training of Call Center Staff

Whether face-to-face or by phone, each interview situation is, by nature, a social interaction between two people. This means that, besides a high degree of professionalism and reliability, a call center interviewer needs to possess excellent social and communication skills. A respondent who perceives the repeated rounds of phone interviews as an unpleasant, boring, or awkward experience is less likely to maintain the effort to answer questions accurately and more likely to drop out of the panel eventually.

High recruitment standards and rigorous training among call center interviewers are therefore vital for the success of an MPPS project. In general, recruitment requirements for field enumerators apply also in the case of call center interviewers, and most of the guidelines for recruiting call center interviewers apply in hiring field enumerators (see below). Box 4.1 gives an idea regarding the number of interviewers needed for the call center.

In addition, a candidate call center interviewer should have the following qualifications:

- Call center experience
- Availability for the duration of the project, to safeguard continuity
- A good telephone manner and ready verbal skills
- The basic information technology awareness necessary to use the data entry interface
- Knowledge of the language of the interview
- Willingness to work late into the evenings and on weekends. For example, in the World Bank's Listening to Africa surveys, call center hours of operation often extend beyond the typical working day: 8 am to 8 pm in the Madagascar and Tanzania mobile phone surveys, 8 am to 5 pm in the Listening to Displaced People Survey in Mali, and 7:30 am to 6 pm in the Togo mobile phone surveys. Some of these call centers operate two shifts. Operating during almost all hours of the day and on weekends, mainly Saturdays, increases the chances of reaching respondents who may be available only at odd times.

The rigorous training of call center staff plays an important role in ensuring data quality and minimizing nonresponse and panel attrition (box 4.2). Typically, call center interviewers do not participate in the full training

---

Box 4.1 The Number of Call Center Interviewers

The number of call center interviewers that needs to be hired depends primarily on the size of the panel, the desired turnaround time, and the number of interviews conducted per day by each interviewer. Experience in the MPPS projects in Madagascar, Malawi, Senegal, Tanzania, and Togo has shown that, during the first few days of a calling round, an average interviewer is able to complete around 15 successful interviews of 15–20 minutes per day. Subsequently, the rate drops to around eight interviews per day, as interviewers start targeting the more challenging respondents, that is, those who could not be reached initially. If we apply these numbers to a concrete example of a target sample size of 1,500, it would take a team of 10 interviewers around 13 days to complete all interviews based on an average estimate of 12 interviews per day.

Box 4.2 Special Considerations in Hiring Call Center Interviewers

In any survey project, sound knowledge of the targeted region is crucial in the creation of questionnaires, planning logistics, and selecting and training interviewers. In conducting mobile phone interviews in Zanzibar, the project leaders of Wasemavyo Wazanzibari, an MPPS project, soon realized the advantages of an all-woman team of interviewers. In the predominantly conservative Muslim society of Zanzibar, men often object if a woman household member is interviewed by a man. Conversely, experience showed that men often preferred to be interviewed by women.

While implementing a mobile phone survey in Gaza, research team members discovered similar gender preferences among respondents. However, it was soon apparent that a small group of mostly elderly respondents were not comfortable talking to a female stranger on the phone. The all-woman team was therefore complemented by a man to allow for some flexibility.

---

program among field enumerators. However, the structure of the training sessions can be organized so that call center interviewers may attend core sessions to familiarize them with the aims of the project and general interviewing techniques.

Even experienced field enumerators should be required to undergo training in conducting mobile phone interviews. In face-to-face encounters, professional enumerators typically rely on nonverbal communication cues such as facial expressions and gestures to put respondents at ease, establish rapport, and defuse tensions. Being sensitive to the nonverbal communication of respondents allows a well-trained enumerator to sense if a question has not been fully understood and detect if the interviewee is becoming annoyed or bored. Because such cues are not available during a phone interview, more stress needs to be put on

Mobile Phone Panel Surveys in Developing Countries • http://dx.doi.org/10.1596/978-1-4648-0904-0

the listening skills of interviewers, on intonation, the pace of speech, and general phone manner.

To plan and structure the training sessions among interviewers, a training manual is useful. Lessons that have been learned during the pretesting of the baseline questionnaires can be fed into this manual.

Incentives and Other Considerations

Various incentives can help ensure the continued motivation and goodwill of respondents throughout the project and can thus safeguard data quality and reduce attrition. An option that has been successfully implemented in MPPS projects is the transfer of phone credits or small cash benefits to cover mobile phone network fees. Because the overwhelming majority of mobile phone users in developing countries use prepaid credit rather than postpaid calling plans, the remote transfer of airtime credit is often a relatively simple and effective way of incentivizing MPPS participation. Many mobile phone providers in developing countries allow client-to-client credit transfers, and it is thus often feasible for interviewers to send airtime credit manually directly after an interview has been completed. However, in large panels, it is advisable to outsource this process to an external party that has a capacity for bulk airtime payments.

While other incentives are conceivable, for example, the transfer of mobile phone vouchers to purchase food at local stores, the remote transfer of mobile phone credit has been applied most widely in MPPS projects. Airtime credits have typically ranged from about $0.50 to $3.00 per round. The exact amount for a project should be determined based on the available budget, the frequency of the survey rounds, and the local cost of phone calls. As a general rule, respondents who are surveyed less often should receive higher per interview compensation because they are more likely to drop out.

An alternative is to offer small payments in mobile money to respondents. In many developing countries, mobile money systems are now widely used even in remote rural areas. This allows for the transfer of small amounts of mobile money that the respondent can then cash at a local agent.

Available empirical evidence on the effect of incentives on MPPS response behavior is mixed. While it is clear that phone credit incentives raise the willingness of respondents to continue participating, experimental data in Latin America and Tanzania show that the value of the airtime credit does not significantly affect the likelihood of respondents dropping out of the panel (Ballivian et al. 2013; Hoogeveen et al. 2014). Evidence from an MPPS project in South Sudan even suggests that attrition is more severe among respondents who have received higher compensation (Demombynes, Gubbins, and Romeo 2013). In contrast, most recent data from a nationally representative MPPS in Tanzania shows that higher incentives promote less attrition (Leo et al. 2015).

Box 4.3 Dynamic Incentives

An innovative approach to incentives has been implemented in Wasemavyo Wazanzibari, an MPPS project in Zanzibar. The project leaders decided to increase the airtime credit incentive after eight mobile phone survey rounds. The rationale was that, once the novelty of the mobile survey has worn off and respondents become accustomed to the balance transfer, the amount may no longer be sufficient. Indeed, after the airtime credit was raised by about $0.20, to $0.63, the participation rates rose in the subsequent round, suggesting this may be an effective way to address short-run drops in response rates.

While there is concern that monetary incentives might not only increase response rates, but actually also influence the answers respondents give, no empirical evidence has emerged that this occurs in surveys in developing countries. For web surveys and household surveys in the United States, studies have shown that incentives do not generally tend to affect the quality or distribution of responses, though they do substantially decrease nonresponse (Singer and Ye 2013) (box 4.3).

Quality Control Measures

Quality control is crucial in any successful call center operation. Devising adequate and effective quality control strategies is therefore an integral element in planning mobile phone survey rounds. The available quality control options depend largely on the software and hardware that are being used. For example, if the call center is relying on professional software solutions, supervisors can listen in on interviews without the knowledge of the interviewers. This not only allows interviewer skills, professionalism, and phone manner to be regularly monitored and enhanced, but also permits supervisors to check that responses are recorded accurately. Alternatively, some call center software packages offer the possibility to record a random subset of interviews that can then be used for quality monitoring later. Other sophisticated systems offer interviewers a bias detection dashboard, that is, a collection of various visual elements arranged on a single screen that shows bias indicators.

If the available infrastructure does not allow for listening in or recording interviews, supervisors might regularly call back respondents to obtain information allowing them to spot-check the performance of interviewers. For this purpose, a small random subset of respondents may be phoned by a supervisor after the regular interview. Callbacks can be used to check if the respondent was indeed interviewed, to assess the respondent's perception of the interviewer's level of professionalism and etiquette, and to repose survey questions to determine if all questions were asked and if responses were entered in the database correctly.

Notes

1. Throughout this document, the term call center is used in a broad sense to refer to a designated space where interviewers are located while they call respondents and enter interview data into the project database using computers, tablets, or other electronic devices.
2. Ideally, a respondent should be contacted by the same interviewer throughout the project (see below in the text).
3. The benefits mentioned here are not exclusive to CATI and may also apply to computer-assisted personal interviewing (CAPI).

References

Ballivian, Amparo, João Pedro Azevedo, William Durbin, Jesus Rios, Johanna Godoy, Christina Borisova, and Sabine Mireille Ntsama. 2013. "Listening to LAC: Using Mobile Phones for High Frequency Data Collection." Final report (February), World Bank, Washington, DC.

Demombynes, Gabriel, Paul Gubbins, and Alessandro Romeo. 2013. "Challenges and Opportunities of Mobile Phone-Based Data Collection: Evidence from South Sudan." Policy Research Working Paper 6321, World Bank, Washington, DC.

Hoogeveen, Johannes, Kevin Croke, Andrew Dabalen, Gabriel Demombynes, and Marcelo Giugale. 2014. "Collecting High Frequency Panel Data in Africa Using Mobile Phone Interviews." *Canadian Journal of Development Studies* 35 (1): 186–207.

Leo, Ben, Robert Morello, Jonathan Mellon, Tiago Peixoto, and Stephen Davenport. 2015. "Do Mobile Phone Surveys Work in Poor Countries?" CGD Working Paper 398 (April), Center for Global Development, Washington, DC.

Singer, Eleanor, and Cong Ye. 2013. "The Use and Effects of Incentives in Surveys." *Annals of the American Academy of Political and Social Science* 645 (1): 112–41.

CHAPTER 5

Conducting Mobile Phone Panel Interviews

Introduction

Mobile phone panel surveys (MPPSs) aim at frequent data collection, typically at least one survey round per month. Each round is a separate activity with a number of key elements that are similar to the building blocks of a traditional field survey, including instrument development, pretesting, training, data collection, data entry, and data cleaning.

This chapter describes the elements of the typical MPPS round, highlighting those aspects that are specific to mobile phone panels. An underlying theme in both the design of a mobile phone survey and the implementation of the call rounds is the minimization of nonresponse and attrition because these reduce the effective size and randomness of the sample.

These terms and the respondent participation decision are examined in the following paragraphs. Subsequent sections outline the core survey elements of questionnaire development, review, pretesting, piloting, training, and data collection. The management of nonresponse and attrition is explained, and empirical evidence deriving from mobile phone surveys is presented on response behavior. The chapter ends with an overview of the main attrition management strategies discussed in the handbook.

Nonresponse, Attrition, and the Survey Participation Decision

Nonresponse refers to the situation arising if a panel member does not participate in a survey round, but continues to be part of the panel. Nonresponse occurs in traditional face-to-face (cross-sectional) surveys, too, for example, if a sampled household can no longer be contacted or refuses to respond. In a phone panel, nonresponse usually means that the respondent cannot be reached on the phone, which may imply refusal.

Attrition refers to the situation arising if a respondent has participated in a survey, but then drops out permanently. Following Alderman et al. (2001), attrition in our case may be defined as the situation arising if a respondent who has participated in the survey at least during the baseline survey stops responding during a call round, r, or verbally confirms a desire no longer to participate.

How is one to distinguish between nonresponse and attrition? If attrition is signaled verbally by the respondent (typically after having been traced), there is no doubt: the decision to discontinue participation must be respected, and the data manager should remove the respondent's number from the call sample and label the respondent among the cases of attrition. Without such explicit communication, if a respondent does not show up in the data for a number of rounds, say four, should he or she be labeled as a case of attrition or nonresponse?

Keeping respondents on the call list if they do not respond retains the option value of their future participation. This benefit needs to be weighed against the cost of the time spent on unanswered calls by the interviewers and by the respondents who decide not to answer and against any problems that might arise because the sample list includes an unrealistic number of active respondents. Any decision will have to consider the possibility of reactivating unresponsive participants against the cost of keeping nonresponding households in the call sample.

Minimizing attrition is critical for at least two reasons. First, even a modest rate of attrition greatly reduces the number of respondents over time and thus the precision or statistical power of survey estimates. Second and more importantly, attrition is likely nonrandom and, if not addressed, adds a bias to the survey results. This has practical implications. High levels of attrition affect the lifespan of the survey, or, inversely, high levels of attrition lead implementers to increase the original sample size. Even small differences in the level of attrition can make a significant difference. A 24-round MPPS aiming to have at least 1,000 responses and with an independent rate of attrition of 1.5 percent per round needs to start with a sample of 1,438 respondents (see chapter 2). However, if the rate of attrition is reduced to 1 percent, the initial sample would only have to be 1,273 respondents, an 11 percent cost saving.

Respondent Decisions to Participate

The decision of a respondent to participate in any survey can be viewed as the outcome of a trade-off whereby respondents weigh costs and benefits. Participation implies a cost in time and mental effort, which may vary with the difficulty, political sensitivity, or embarrassing nature of the questions. For a positive participation decision, this cost has to be offset by (1) the benefit of an incentive received after completing the interview, plus, possibly, (2) the intrinsic benefit of participating. This type of decision making has been modeled by Hill and Willis (2001), who argue that the rewards of survey participation may include monetary payments and intangible rewards that can be influenced by the design and implementation of the survey. The latter might encompass the benefit of being consulted and thereby contributing to the welfare of the nation and the benefit of having an interesting conversation that provides food for thought.

Another intangible aspect affecting the comfort of the respondent is the random selection among other household members to represent the household and to talk to outsiders; respondents may consider this a cost or a benefit. Trust is also an important aspect. Respondents should be confident their data are used only for statistical research and are not associated with personal identifiers. The quality of the interviewer is likewise crucial. Hill and Willis (2001) report that contact with the same interviewer over multiple rounds has a strong positive impact on response rates. Lugtig (2014) adds that the propensity to respond to a survey is positively affected by commitment to the survey, habit formation, and incentives. Conversely, panel fatigue and negative shocks tend to lower the propensity to respond.

Groves, Cialdini, and Couper (1992) use a typology of factors affecting survey participation that distinguishes societal factors (survey density); survey design, including mode of (first) contact, incentives, survey protocols, and questionnaire length; characteristics of the respondent, for example, the socioeconomic and demographic profile; and interviewer attributes, such as age and gender, but also interviewer personality and confidence. Lynn et al. (2005) review the respondent cost-benefit analysis of survey participation and emphasize the importance of reciprocity in the interview process, whereby survey managers attempt to establish trust and comfort and to use incentives to nudge interviewees. They agree with earlier literature that the (first contact) interviewer-interviewee interaction is important, as are the social value of the survey and the prominence of the survey topic.

In a mobile phone survey, the participation cost-benefit calculation is arguably different relative to an interview by an unannounced enumerator on the household doorstep. On the positive side, the payment of the incentive, conditional on participation, is clearly defined before the phone survey interview. Moreover, the baseline survey, with its consent form, joint-identification of a suitable time for follow-up, phone distribution, and respondent group meeting, already likely represents a level of commitment that may be cemented by frequent conversations with the same call center interviewer. Thus, the MPPS model described in this book is likely to create commitment and habit formation, and, through careful survey preparation and enumerator and interviewer training, respondent commitment can be enhanced.

An additional element in the maintenance of survey commitment is a periodic respondent evaluation round in which respondents are asked about their views of the survey and are invited to provide feedback. This allows survey managers to take stock of self-reported fatigue and consider suggestions by respondents, for example, suggestions on survey round themes. Feedback on the use of and media reporting on survey results may also provide an incentive to respondents.

On the negative side, many respondents may find that a refusal to participate expressed simply by not picking up the phone is easier than saying no to an enumerator on the doorstep. Moreover, respondents likely experience panel fatigue, particularly if survey rounds are frequent; interviews are long; and the questions are difficult to answer. Nonresponse will clearly remain a fact of life in MPPSs, but it may be reduced through careful preparation.

Mobile Phone Panel Surveys in Developing Countries • http://dx.doi.org/10.1596/978-1-4648-0904-0

The Mobile Phone Survey Questionnaire

Questionnaire Development

A specialist literature on survey design and questionnaire development has emerged based on decades of household survey experience and data analysis. In their Living Standards Measurement Study (LSMS) handbook, Grosh and Glewwe (2000) focus on questionnaire development in multitopic household surveys. However, many of their guidelines are equally relevant here since, in essence, mobile phone surveys may be considered a special type of multitopic surveys wherein the various thematic modules are spread across a number of subsequent interview sessions (the call rounds), with longer lag times between sessions than in face-to-face interviews. Mobile phone surveys might even cover all LSMS questionnaire types: household questionnaire, community questionnaire, and price questionnaire.

The latter two types of questionnaires can, in principle, be administered fairly easily by calling designated respondents, particularly if these are enlisted as community respondents or monitors during survey preparation. The independent and frequent monitoring of the prices of a basket of goods by a designated monitor is an extension of mobile phone surveys and address an often-heard complaint about downward bias in official inflation data. School and health care facility interviews are possible as well.

The design of questionnaires is at the heart of traditional survey design. Grosh and Glewwe (2000) distinguish five steps in the design of multipurpose surveys: formulation of the overall objectives of the survey, choosing which modules to include in the questionnaire and the length of each, the design of the individual modules, the integration of modules into one questionnaire, and translation and field testing.

Step 1, the formulation of the overall objectives of the MPPS, is addressed in chapter 1. In step 2, the length of individual modules is a critical variable in the design of mobile phone survey rounds (see below). The choice of modules or questions within modules (step 3) is less prescriptive in panel rounds than in traditional field surveys because new modules or questions can be added at low cost. However, defining a yearly calendar that schedules all modules is important so that priorities can be set in light of the overall objectives.

At the same time, a unique strength of mobile phone surveys is the ability to change priorities and calendars, that is, their flexibility (chapter 1). For example, if a communicable disease triggers a national health crisis (such as the Ebola crisis), policy makers need representative data about citizen experiences, knowledge, and practice immediately. Similarly, if the release of examination data leads to a national debate about the state of education (such as in Tanzania), the ready availability of recent data on teacher absences is likely to have much more political value for policy makers than less recent data.

Step 4, the overall survey integration of individual modules, is less relevant for the case of mobile phone surveys. Step 5, translation and testing, the last step, is addressed in the next section. The remainder of this section highlights general recommendations on questionnaire development from the literature. Appendix D illustrates a sample mobile phone survey questionnaire. The aim is to emphasize the aspects that are specific to these surveys.

The Elements and Content of the Questionnaire

A typical mobile phone round questionnaire has the following elements. First, respondent verification: after making contact, the call center interviewer uses baseline data on name, age, and sex to assess whether the person answering is the originally sampled respondent. Because a share of the adult population is expected to migrate, these questions should periodically include an inquiry about the present location of the respondent. After respondent identity confirmation, the call center interviewer introduces the topic. Questionnaires may be structured into repeated question sections and new question sections. A final fixed part of the questionnaire might allow for interviewer observations related to survey management: How many calls were used to reach this respondent? Was the respondent reached directly or through tracing, whereby contact with the respondent was attempted through a family member, the group leader, or another respondent? Who was used for tracing?

An iterative process of questionnaire development is recommended. As many relevant people as possible ought to be involved, particularly if location-specific factors and policy questions are important. If the mobile phone survey covers a broad range of themes, the expertise of a broad group of people may be required, as in the case of multipurpose household surveys.

The value of new, location-specific questions needs to be compared with the value of comparability across surveys, which is an advantage. Often, questionnaire development is not launched in a void, but makes use of existing questionnaires. This facilitates the comparison of final estimates with findings on other countries or locations or on earlier time periods. Within-country comparability over time is a particularly significant aspect of mobile phone surveys, which, because of the high frequency of these surveys, allows the monitoring of indicators. Indeed, to permit trend analysis of the same indicators, it makes sense not to change the wording of questions in the baseline survey that are used in the MPPS.

If the wording and nature of questions are maintained in the questionnaires of other surveys, this improves comparability across surveys and, possibly, across countries. A good example of this standardization is provided by "Core Questions on Drinking-Water and Sanitation for Household Surveys" (WHO and UNICEF 2006). The use of these standardized questions allows for precise comparisons, such as "we observe that country X has improved access to clean water from 50 percent to 60 percent, but neighboring country Y has remained at 40 percent over the same period." Such comparisons enhance the survey value for policy makers.

Some topics are more well suited for use in mobile phone surveys than others. Asking for opinions is relatively easy on the phone: "What is the main problem facing your country?" or "If elections were held today, which candidate would you vote for?" Likewise, questions about social service delivery indicators can readily be asked over the phone: "Did your child go to school yesterday?" or "Did your child sleep under an insecticide-treated mosquito net last night?" It is difficult to obtain precise anthropometric measurements during a traditional field survey with trained enumerators; noisy measurement may seriously reduce the value of the exercise especially if the absolute differences between units or over time are small. This seems even more daunting if the discussion with a household takes place over the phone unless child health cards with recent anthropometrics are available. Piloting the questionnaire should enable the manager to decide on the feasibility of adding relevant questions.

Some types of data collection are obviously unsuitable for use over the phone, for instance, questions needing special tools such as scales. Because of their length, LSMS-type consumption expenditure and production modules are a challenge for enumerators and households even in face-to-face situations; they are probably too long for a phone interview. Conversely, keeping questions short and simple, preferably without many answer categories that need to be read out, is a safe choice. Between these extremes, it is not easy to give hard and fast rules. Filling out a household roster over the phone is not impossible, but there is a strong case for doing these during the face-to-face baseline survey because they take a lot of time, and it is helpful to have visual contact with at least part of the roster for verification. However, a household roster that has been collected at baseline can be verified over the phone a year later to assess demographic changes.

It is not true that complexity should strictly be avoided. Relatively complex questions have been successfully fielded over the phone, for example, questions on stocks, dates of expiry, and pricing for a range of essential medicines in health facilities. A prerequisite is that the questions be well structured and easily understood by the call center interviewer and respondent, which depends on a good scripting phase. Dillon (2012) used picture cards in his Tanzania phone survey. He left a laminated set of cards connected by a ring with each respondent, and, during phone interviews, the respondents were asked to turn to the appropriate page where they could view pictures that helped them respond to subjective probability questions.

Food consumption data, such as those proposed by the World Food Programme for use in food security analysis, can be collected through mobile phones (WFP 2015). The frequency-weighted diet diversity score, for instance, is calculated using the frequency of consumption of various food groups by households during the seven days previous to a survey. The World Bank's Listening to Africa surveys collect these data on a regular basis because they are useful in monitoring vulnerability and food security (see appendix D).

The accuracy of responses can be significantly enhanced through the careful design of questionnaire skip patterns. Iarossi (2006) notes that respondents rarely admit ignorance and tend to answer any question. Target respondents are typically sampled randomly, and one should expect that not all questions apply to all respondents equally. For example, a question on the reception by farmers of fertilizer vouchers should typically follow a question establishing whether or not the respondent is a farmer. This ensures that questions are relevant and avoids irritating the respondent. Moreover, questions should adhere to the BOSS rule (brief, objective, simple, and specific), that is, be brief (avoid joining multiple questions into one), objective (avoid suggesting an answer in the question and making assumptions about the respondent), simple (avoid jargon, acronyms, double negatives), and specific (avoid unspecific or vague descriptions and ask about specific events and periods).[1]

Duration and Frequency of Mobile Phone Surveys

As a rule of thumb, interviews during mobile phone surveys are each restricted to about 15–30 minutes, and a typical questionnaire contains around 20 questions. The suggested duration of an interview is based on common practice in recent household surveys relying on mobile phones. There is no other supporting evidence. Researchers might therefore investigate the optimal length of the interview.

Qualitative assessments by call center supervisors indicate that lengthy questionnaires lead to fatigue and reduced cooperation among respondents. In an evaluation of the Sauti za Wananchi MPPS in Tanzania by respondents, however, only 4 percent of the respondents suggested that fewer questions would enhance the survey, against 29 percent who felt that offering better incentives would improve the initiative.[2] Iarossi (2006) cites literature that finds a weak association between questionnaire length and response rates. There is, however, some empirical evidence that an increase in questionnaire length is associated with a decline in response rates and the quality of the responses to phone surveys (McCarty et al. 2006; Roberts et al. 2010). In any case, in mobile phone surveys implemented thus far, the questionnaires have typically been much shorter than the average questionnaires in multipurpose household surveys.

Given a fixed number of survey questions to be asked over a year, any decision on the average module length also determines the frequency of calls and vice versa. However, a typical call center contract will specify a fixed cost per round and limits on the number of minutes per call. Thus, the survey budget is likely to dictate the number of rounds and, to some extent, the length of calls.

A core advantage of mobile phone surveys is that they facilitate the frequent monitoring of indicators, for example, the availability of clean water among households. Effective monitoring requires that the same questions be asked at multiple points in the life of a survey. However, based on respondent evaluation surveys and discussions with call center interviewers, there is anecdotal evidence that respondents become annoyed if the same survey topic is repeated in two consecutive rounds. Keeping some variation in survey topics over consecutive rounds is therefore recommended to prevent respondent fatigue. As a rule of thumb, mobile phone surveys avoid repeating topics and questions more often than every six months. Unless the implementation of a specific policy requires scrutiny, a road building project, for instance, or indicators that may vary widely over time, such as food prices during food crises or security during conflict, this frequency seems sufficient to follow trends accurately.

Many socioeconomic variables typically measured in surveys change only slowly; access to improved water sources is a good example. To reveal a statistically significant change between survey rounds in such a case would demand a long wait. Yet, the lack of change is an important survey result as well, particularly if public expenditure is to be assigned to programs to enhance the performance measured by statistical indicators. Opinions on issues that are being debated prominently in the media can be expected to be more variable; a good example are the changing views on proposals to revise the Constitution of Tanzania.

Review, Translation, Pretest, Scripting, Training, and Piloting

Once a final draft questionnaire has been compiled after an iterative process of inputs and review, five steps follow. First, the project manager reviews the questionnaire for consistency, clarity, and flow. Because a mobile phone survey questionnaire is much shorter than a standard multipurpose household questionnaire, it typically consists of the equivalent or less of a thematic module in such a questionnaire. This means the review can, within a short time, reveal whether all questions are clear, in logical order, and associated with appropriate, understandable answer categories and whether all skip patterns are correctly placed. Particular attention must be paid to question instructions on whether to read out answer options. If the budget allows, the questionnaire is pretested among a small set of out-of-sample respondents. Alternatively, the call center interviewers can perform mock interviews to validate the questionnaire prior to commencement of the phone interviews.

At this stage, the questionnaire will normally still be in text format awaiting data entry to the call center software. The project manager or lead enumerators implement the pretest and take notes during the pretesting interviews. In addition, input provided by respondents during and after the interview can be a valuable resource for revising and improving the questionnaire. Lessons learned during the pretesting phase are used as input for the training. Once the review and pretest are completed, the questionnaire is in near final form on paper.

If required, the next step is the translation of the questionnaire into local languages. This may appear to be a minor detail, but it is not. Particularly if theme-specific jargon is used, such as in surveys on capitation grants, improved water sources, or subsidized irrigation schemes, translations are nontrivial. A good technical translator is a major asset.

The subsequent step involves scripting the questionnaire, that is, translating the questionnaire into a file that allows for easy data entry.[3] In a standard field survey, an equivalent process involves programming a data entry mask reflecting the questionnaire into software, for example, for a handheld device or for a data entry program. Such programming typically also includes the skip patterns and can check for data entry errors, which can save time after data entry.

When the scripted questionnaires are ready, the call team can be trained. Field surveys typically parcel out the labor. Interviewer teams might do the interviewing, while data clerks do all the entry. In a mobile phone survey, these two functions are performed by one person, who, typically, is working in a team.

A procedure whereby a call team sits in close proximity to a supervisor offers many advantages during the call process. A good supervisor will be able to listen in and pick up any issues that need to be clarified during piloting and even during the first survey calls if necessary. For this reason, the supervisor needs to lead the training session, be the champion of each round's questionnaire, and know it inside out. During the review process, the survey manager and the call supervisor need to have worked together closely to flag any question phrases that are unclear. Potential issues need to be raised and clarified during the next pilot phase. The training itself might consist of a close reading of the questionnaire, question-by-question, so that a shared understanding of the contents emerges. Call center interviewers should provide comments and be able to answer any question that may arise among respondents.

Immediately after the training session, the piloting of the questionnaire by different team members takes place. We distinguish a pilot process from a pretest; whereas a pretest only seeks to determine if the quality of the questionnaire is good, a pilot procedure tests all the elements of the data collection process, including the technology, the scripted questionnaires, and data entry. Ideally, the pilot procedure focuses on respondents who are outside the survey sample, but who otherwise receive the same treatment. Any final questions should be addressed and answered by the supervisor and the survey manager at this stage.

Data Collection

The Collection Process

Relative to face-to-face field surveys, the logistics of the individual rounds of the MPPSs are simple. Once the questionnaires have been scripted and the training and piloting have been completed, the call center interviewers may begin calling respondents.

Because the baseline has been completed by this stage, there is no need for sampling. However, during each call round, there may be minor changes in the round sample or call list because of attrition. Therefore, between call rounds, changes in the call lists are reviewed based on the results of the previous call round. Before the call round begins, the data manager prepares the database by loading all the nondropped respondents into data folders for each call center interviewer. If substitute respondents are available in the enumeration area (EA) where respondents have dropped out, they are added to the call list.

The frequency of calls in a year determines the interval at which panel members are called. The frequency decision is guided primarily by existing data needs. In addition, it is important that the panels do not remain inactive too long because respondents tend to lose interest and are more likely to drop out if they are not approached regularly. Nonetheless, calling too often may create irritation and fatigue among respondents, thereby seriously affecting data quality and attrition rates negatively.

In most settings, this trade-off means that respondents should be approached no more than twice a month, although some surveys, such as Listening to Dar, have relied on weekly rounds. If a panel of respondents is left alone for more than three months, respondents are likely to break out of the survey routine and become less committed to the survey. Mobile phone surveys now typically aim for a high frequency of at least one call per respondent per month. However, there is no clear evidence of the impact of frequency on response rates.

To minimize nonresponse and attrition, phone calls should occur at regular intervals and within the same time window, for example, on weekends or in the evening. Ideally, respondents have been asked during the baseline interview to indicate the best time of the day and week to reach them, as well as their preferred phone number.

Experience has shown that, to build up the goodwill and commitment of respondents with respect to the project, respondents ought to be contacted by the same call center interviewer as much as possible. This allows the respondent to become used to the voice of the interviewer and renders the interviewing experience more predictable and personal. The personal relationship adds to the formation of a routine in survey participation. Some respondents reportedly look forward to the regular phone conversations.

The call calendar depends on the sample size. Experience teaches that call interviewers can complete between 10 and 15 interviews successfully per day (see box 4.1, chapter 4). Thus, for a sample of 2,000 households, a team of 10 call center interviewers making an average of 12 successful calls per day would need a call period of about 17 call days. In practice, however, the daily productivity of interviewers is not constant over the duration of a round. Figure 5.1 provides a typical illustration: the first week of calls yields the low hanging fruit and

Figure 5.1 Completed Interviews, by Day, 2013 Sauti za Wananchi Survey, Round 1

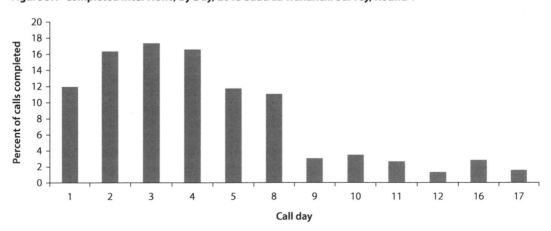

Source: Twaweza, Sauti za Wananchi mobile phone survey, 2013, round 1.

accounts for more than 70 percent of the total interviews. The yield is much lower in week 2, but this is also a reflection of the fact that the call center worked one shift instead of the initial two to complete the remaining 30 percent of the total interviews. This approach and yield pattern is maintained in later rounds; thus, the difficult task of reaching the remaining 20–30 percent of respondents is dealt with by a smaller dedicated team. A sample phone survey round calendar is presented in appendix E.

Quality control during collection is achieved primarily by the call team supervisor, who listens in on interviews and can intervene or correct on the spot. Supervisors need not listen to all interviews, but can systematically listen to a certain proportion of the interviews at random, for example, 5 percent. This is only an additional quality control measure and does not replace other measures, including monitoring the dashboard of the results to prevent data entry errors. A good supervisor detects team members who need support and acts accordingly. A second layer of quality control is represented by the data manager, who can run preliminary data range checks during the collection period and ask for clarification. In case of severe problems, it is not difficult to return to a household and reinterview in a mobile phone survey. Quality control can also be accomplished between rounds by monitoring individual interview quality based on the data collected and on the recorded calls. If calls are not recorded, the supervisor might call a subsample of respondents and verify survey participation. Another type of verification involves calling a random subsample of the respondents who were not initially reached by the call center.

The Challenges in Data Collection

For each call that a call center interviewer makes, there are four possible basic outcomes, which are listed as follows in decreasing order of success:

1. The respondent is reached through a direct call to his or her phone, and the interview is completed.
2. The respondent is not reached directly, but through a tracer, and the interview is completed.
3. The respondent is not reached in the current round, but is retained in the sample.
4. The respondent drops out of the panel (attrition).

In this subsection, we describe practices that deal with these scenarios. For the eight call rounds completed in the 2014 Sauti za Wananchi survey, figure 5.2 provides evidence on the relative importance of these scenarios. The data only describe eventually successful calls. In nearly all rounds, scenario 1 plays out in at least 90 percent of the calls, and the respondent is reached directly on his or her phone.

Even if respondents are reached directly, the call center interviewers may have to call them more than once. Figure 5.2 shows, however, that the large majority of interviews are completed during the first direct calls to respondents. This is an important finding. The Sauti za Wananchi panel had been interviewed over 12

Figure 5.2 2014 Sauti za Wananchi Survey, Interviewers: How Did You Reach the Respondent?

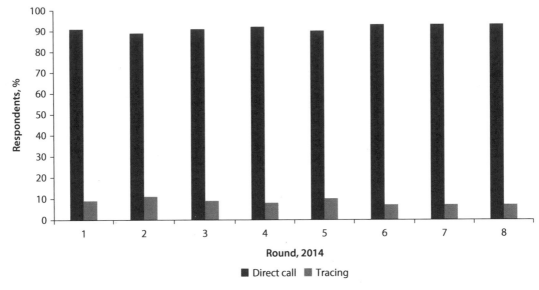

Source: Twaweza, Sauti za Wananchi mobile phone survey, 2014, rounds 1–8.

rounds in 2013. In early 2014, the response rates were at about 80 percent of the original baseline sample (see the next subsection). Figures 5.2 and 5.3 show that these active respondents were, in most cases, reached easily, with only one direct call from the call center to the respondent number.

What if a respondent has not been reached after several direct calls? There are many reasons why a respondent may prove difficult to reach. In the 2014 Sauti za Wananchi survey call rounds, respondents who were difficult to reach, but were eventually traced through a third person, were asked why it took so long to reach them. Figure 5.4 shows the main reasons for the delay among those respondents who could not be reached through a direct call. The figure shows there is some persistence in the distribution of the answers and that respondents also highlight technical issues such as low phone battery, recharging, and network problems.

These challenges appear to be common. In the World Bank Listening to Africa Project, a review of the reasons some households in Madagascar could not be reached yielded the following:

- The phone had been turned off because household members were not used to possessing a phone.
- Improper use of the phone or the SIM card.
- Lack of network coverage.
- Technical problems with the phone.
- Lost or stolen phone.
- Refusal to continue participation in the survey.
- Lack of sufficient electricity or sunlight to recharge the phone.

Figure 5.3 2014 Sauti za Wananchi Survey, Interviewers: How Many Times Did You Call This Respondent?

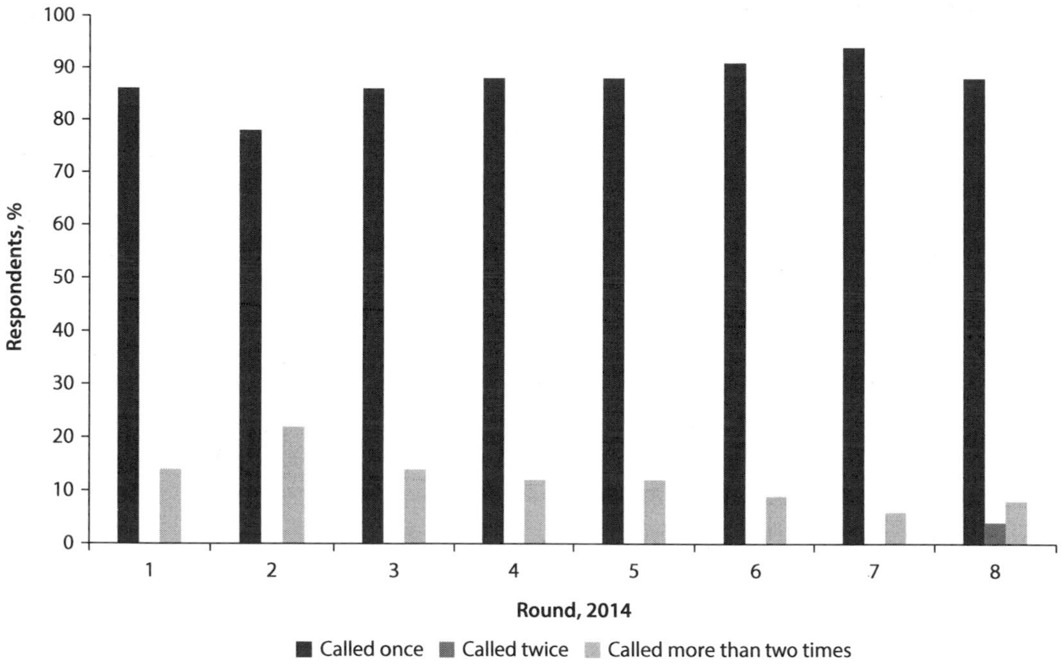

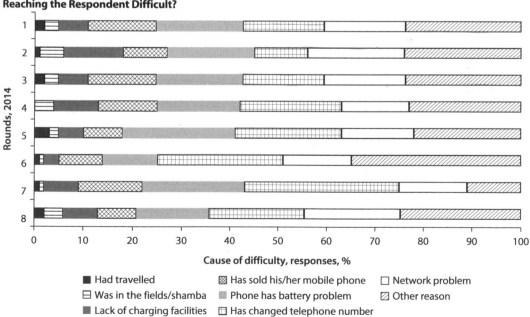

Source: Twaweza, Sauti za Wananchi mobile phone survey, 2014, rounds 1–8.

Figure 5.4 Respondents Who Were Traced, 2014 Sauti za Wananchi Survey, Interviewers: Why Was Reaching the Respondent Difficult?

Source: Twaweza, Sauti za Wananchi mobile phone survey, 2014, rounds 1–8.

The design of a mobile phone sample should involve strategies to address the problems of nonresponse and attrition. First, the commitment of respondents to the survey should be instituted and documented during the baseline study as much as possible (see chapter 3). Second, alternative phone numbers should also be collected on each respondent during the baseline. Then, if the primary or preferred respondent phone number does not respond, the call center may try the alternative numbers. Third, the respondent might be traced, that is, a family member, the relevant group leader, or another respondent in the same EA is asked to make contact with the respondent (chapter 3). Tracing is not always successful, but it can account for some 10 percent of the final dataset (see figure 5.2). The tracing strategies used by the Sauti za Wananchi survey call center are illustrated in figure 5.5.

Tracing is more difficult than direct calling, and call center managers have complained about the problems associated with seeking the help of group leaders and other respondents in tracing target respondents. However, 90 percent of the respondents in the Tanzanian survey indicated they were willing to help in tracing.[4]

The strategies used in mobile phone surveys may include, but are not limited to the following:

- Maintaining a regular call schedule for each respondent.
- Calling the personal phone numbers and alternate phone numbers of each respondent or household.

Figure 5.5 Tracing Strategies Used to Reach Respondents, 2014 Sauti za Wananchi Survey

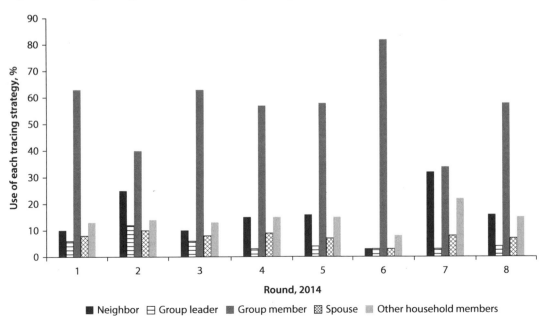

Source: Twaweza, Sauti za Wananchi mobile phone survey, 2014, rounds 1–8.

- Calling neighbors, family members, paired households, or community leaders to establish contact with the respondent or household.
- Calling the respondent or household early in the morning, late at night, or on weekends.
- Sending an SMS to the respondent or household and, upon confirmation of message delivery indicating that the phone is on and covered by the network, making the call.
- Use of messengers in the EA to establish contact with the respondent or household.
- Calling as many times as possible during the survey round to reach respondents or households that are difficult to reach and, unless they have dropped out of the panel, calling again in later months those who did not participate in previous months.

Response Correlates and the Effect of Incentives

This section provides evidence on the correlates of response behavior. The small but growing literature on high-frequency mobile phone surveys provides some evidence on respondent characteristics that may be used to predict the number of successful calls. Relevant surveys are listed in the top row of table 5.1. These surveys are quite different in context, but share typical MPPS design aspects.[5] Here, the issue is to determine the findings that are common in the various settings through the presentation of the main regression results qualitatively.

Table 5.1 Effect of Various Variables on Phone Survey Response Rates

Variable	South Sudan, urban	Dar es Salaam, Tanzania	Honduras	Tanzania mainland
Baseline sample size	1,007	550	1,500	2,000
Phones distributed	To all	No	Respondents w/o phones	To all
Age	+	0	+	+
Household had a phone prior to baseline survey	+	+	+	+
Men	+	0	0	+
Rural		−	0	0
Wealth		+	+	+
Education		0	+	+
Network strength, operator indicator	+	+	+	+
Design variables				
Airtime incentive, randomly assigned	−	0	+	+

Source: Summary based on the multivariate regression analyses reported in Ballivian et al. 2013 (Honduras); Demombynes, Gubbins, and Romeo 2013 (urban South Sudan); Hoogeveen et al. 2014 (Dar es Salaam, Tanzania); Schipper et al. 2014 (Tanzania mainland).
Note: Cell contents signs +/0/− mean positive significant/not significant/negative significant, respectively.

The strongest common findings are that respondents tend to participate in survey rounds more often if they (1) are living in households in which a member is already using or is familiar with mobile phones, (2) use particular network providers, and (3) are living in more well-off households. The first finding signals that phone distribution, at least among households without phones, is a must in MPPS design, but it is clearly not a sufficient condition for survey participation. The second factor relates to network strength, which differs across providers and suggests that the baseline needs to screen rigorously for network reception at the dwelling and to either drop households and EAs in which reception is poor or find another means to communicate with these households. The finding on wealth suggests that oversampling among poor households or the targeting of incentives based on poverty levels may be useful.

Respondent age, sex, urban or rural location, and formal educational attainment are relatively weaker overall predictors of survey participation. Apparently, these factors are context specific, and the findings do not provide clear, relevant lessons for survey design.

Incentive levels are among the key survey design features. They have a direct impact on the survey budget, but also on the cost-benefit calculation of the respondent. In settings in which trust is lacking, paying the incentives in a predictable manner is a powerful signal that the survey organization is serious, and this may stimulate commitment.

Incentives are also an ideal design feature for analysis because they are easy to randomize. A standard practice is to determine the amount of the incentives and then randomly assign them within EAs to prevent questions or conflicts within communities. The evidence appears to be mixed (see table 5.1). Particularly puzzling is the South Sudan finding, which shows a negative incentive effect. Thus, the value of the incentive offered was negatively correlated with survey completion; participants who were offered the SD 10 airtime credit were about 6 percent less likely to complete the survey (Demombynes, Gubbins, and Romeo 2013).

The other studies found either a zero or a positive effect. The direction of effect sizes within the Honduras survey and the Sauti za Wananchi survey in Tanzania consistently conform to expectations: higher incentives result in higher response levels (Ballivian et al. 2013; Twaweza 2013). In the Listening to Latin America and the Caribbean samples, three levels of incentives were provided: zero, $1.00, and $5.00. The impact of the incentives differed across contexts. In Peru, after six rounds, attrition was higher among the zero incentive group, but no difference was observed between the low- and the high-incentive group. In other words, it seems the level of the incentive did not matter. The results in Honduras are different. There, the endline response rates are barely different between the zero and the low incentive, while they are higher for the high incentive. While these settings differ in the details of the findings, two common elements are that (1) incentives do seem to trigger differential response behavior, and (2) the direction of the impact is consistently as expected in that higher incentives lead to better response rates. The reason for the zero effect in the Dar es Salaam case may well be that both the sample size and the incentive level differences were relatively small.

We conclude that airtime incentives need to be in place to increase the trust and loyalty associated with the survey. The exact level of the maximum incentive is likely to be bounded by the budget and the notion that the extent of the intervention of the survey should be minimized. In any case, incentive levels appear to affect response rates and may be instrumental in maintaining the responsiveness of the sample. In the Tanzania mainland survey, the incentive levels were raised after one year, and this had a positive impact on response rates.

Attrition Management: A Review

This last section presents a summary of all the MPPS attrition strategies described thus far. We present these strategies in table 5.2 in the approximate chronological order of MPPS project implementation. Tables 5.2 and 5.3 present attrition management strategies during the baseline and phone surveys, respectively.

The success or failure of an MPPS depends greatly on the ability to maintain an active and robust panel throughout the lifetime of the survey, that is, the

Table 5.2 Attrition Management, Phase 1: Baseline Survey

Strategy	Rationale
Sampling: sample EAs with good average network reception; sample households with good reception at dwelling	Respondents should be reachable on their mobile phones at least part of the day
Hardware: providing mobile phone, SIM card (preregistered), and charging solutions to respondents; at the end of the baseline interview, the enumerator calls the call center using the respondent's phone	Lack of phone ownership or access to electricity should not be a barrier to sample inclusion; make sure the technology is functioning to ensure future participation
Consent: use forms to document both respondent consent and head of household consent, signed	Creates commitment to the survey, documents the participation agreement; the household head consent form makes sure the survey initiative has been explained to the head of household and may prevent domestic conflicts later
Data collection: collecting alternative numbers: preferred daytime and nighttime phone numbers, contacts of household members who own mobile phones; collecting information on the time that each respondent prefers to be called; recording the contact information of relevant people outside the survey (neighbors, contacts)	To make sure respondents can either be reached or traced through family, friends, and neighbors
Respondent groups: form respondent groups, select group leaders, carry out group training; during the group meetings, encourage teamwork in tracing respondents; ask about the willingness of the group leaders to conduct tracing and act as village monitors	Respondents motivated by group membership; saves time in explanation, phone and charger instruction Group leaders can be helpful in tracing respondents; they can act as local survey representatives and as monitors of local service provision, prices, and so on

table continues next page

Table 5.2 Attrition Management, Phase 1: Baseline Survey *(continued)*

Strategy	*Rationale*
Sample reserve respondents, either among phone owners or by providing phones	To replace respondents who drop out
Preliminary call rounds: start calling respondents soon after the household visit, during the finalization of baseline fieldwork	To keep in touch; confirm or establish trust, follow up on baseline promises, and prevent early attrition before call rounds
Communication, community information: through community entry meetings by field teams or through printed booklets	Explaining the survey to community leaders and members helps in implementation, encourages acceptance, and helps avoid conflicts; emphasis on the lottery nature of random sample selection
Top respondent award: a gift, for example, a radio, to be provided after two years to the 100 most consistent respondents	Creates better response rates

Table 5.3 Attrition Management, Phase 2: MPPS

Strategy	*Rationale*
Call center: Ensuring the call center is equipped with the proper technology and an experienced call center team; provide incentives for top interviewers and good data managers	An efficient call center and data operation will help enormously
Incentives: provide respondents with an incentive at the end of each MPPS data collection round	A fixed monetary reward for each round maintains support and response rates
Interviews: start calling immediately after baseline, regularly; limiting the length of the telephone interviews; avoiding calling respondents outside of preferred calling times	Regular contact to maintain respondent loyalty is particularly important immediately after the baseline phase; helps avoid respondent fatigue
Matching of call center interviewers and respondents: having the same interviewer contact a given respondent for each survey to foster respondent and interviewer understanding; ask about and respect interviewer gender preferences of respondents to the extent possible	Maintain trust and comfort among the respondents
Tracing: using other people in the community to follow up on respondents who cannot be reached over the phone	Raises the response rate
Call center interviewer behavior: thank the respondents for participation; always speak politely and appreciatively	Basic courtesy; limits fatigue
Respondent feedback, visit: consider visiting the respondents after every year of calling; provide qualitative feedback on reports, use of the data, media coverage	This may keep the respondent sample engaged and improve intrinsic motivation

maintenance of high response rates and low rates of attrition. These challenges are not unique to mobile phone–based panel surveys, though mobile phones may provide unique challenges and opportunities in managing attrition.

The implementers of an MPPS must therefore carefully analyze the causes of attrition and identify relevant opposing strategies. This section outlines the strategies that have been used in the past with apparent success. However, in most cases, there is no rigorous evidence on the impact these strategies might have; the correlations are descriptive at best. Rigorous evidence exists only in the case of the impacts of airtime incentives.

An MPPS might be associated with an endline survey. At the start of the MPPS, a time horizon for the endline survey might be provided so as to be transparent with respondents. An endline survey represents an opportunity to provide and receive any final feedback on the survey and to show a token of appreciation to all participating households. An ending date for survey commitment should also be laid out clearly. This also clarifies the date when ownership of the phone and charger is transferred to the respondents.

Notes

1. See Iarossi (2006) for details and examples.
2. Data compiled from the Sauti za Wananchi respondent evaluation round, November 2014.
3. Seeking advice on the appropriate translation of key terms may be advisable especially because there is scope for misunderstanding, which may be greater in a telephone interview.
4. Data compiled from the Sauti za Wananchi mobile phone survey, evaluation round, January 2014.
5. Note, however, that the Honduras survey results include data on phone calls, SMS, and interactive voice response (IVR). Please refer to the papers listed in the table source line for details.

References

Alderman, Harold, Jere R. Behrman, Hans-Peter Kohler, John A. Maluccio, and Susan Cotts Watkins. 2001. "Attrition in Longitudinal Household Survey Data." *Demographic Research* 5 (4): 79–124.

Ballivian, Amparo, Joâo Pedro Azevedo, William Durbin, Jesus Rios, Johanna Godoy, Christina Borisova, and Sabine Mireille Ntsama. 2013. "Listening to LAC: Using Mobile Phones for High Frequency Data Collection." Final report (February), World Bank, Washington, DC.

Demombynes, Gabriel, Paul Gubbins, and Alessandro Romeo. 2013. "Challenges and Opportunities of Mobile Phone-Based Data Collection: Evidence from South Sudan." Policy Research Working Paper 6321, World Bank, Washington, DC.

Dillon, Brian. 2012. "Using Mobile Phones to Collect Panel Data in Developing Countries." *Journal of International Development* 24 (4): 518–27.

Grosh, Margaret, and Paul Glewwe, eds. 2000. *Designing Household Survey Questionnaires for Developing Countries: Lessons from 15 Years of the Living Standards Measurement Study.* 3 vols. Washington, DC: World Bank.

Groves, Robert M., Robert B. Cialdini, and Mick P. Couper. 1992. "Understanding the Decision to Participate in a Survey." *Public Opinion Quarterly* 56 (4): 475–95.

Hill, Daniel H., and Robert J. Willis. 2001. "Reducing Panel Attrition: A Search for Effective Policy Instruments." *Journal of Human Resources* 36 (3): 416–38.

Hoogeveen, Johannes, Kevin Croke, Andrew Dabalen, Gabriel Demombynes, and Marcelo Giugale. 2014. "Collecting High Frequency Panel Data in Africa Using Mobile Phone Interviews." *Canadian Journal of Development Studies* 35 (1): 186–207.

Iarossi, Giuseppe. 2006. *The Power of Survey Design: A User's Guide for Managing Surveys, Interpreting Results, and Influencing Respondents*. Washington, DC: World Bank.

Lugtig, Peter. 2014. "Panel Attrition: Separating Stayers, Fast Attriters, Gradual Attriters, and Lurkers." *Sociological Methods & Research* 43 (4): 699–723.

Lynn, Peter, Nicholas Buck, Jonathan Burton, Annette Jäckle, and Heather Laurie. 2005. "A Review of Methodological Research Pertinent to Longitudinal Survey Design and Data Collection." ISER Working Paper 2005–29, Institute for Social and Economic Research, University of Essex, Colchester, United Kingdom.

McCarty, Christopher, Mark House, Jeffrey Harman, and Scott Richards. 2006. "Effort in Phone Survey Response Rates: The Effects of Vendor and Client-Controlled Factors." *Field Methods* 18 (2): 172–88.

Roberts, Caroline, Gillian Eva, Nick Allum, and Peter Lynn. 2010. "Data Quality in Telephone Surveys and the Effect of Questionnaire Length: A Crossnational Experiment." ISER Working Paper 2010–36, Institute for Social and Economic Research, University of Essex, Colchester, United Kingdom.

Schipper, Youdi, Elvis Mushi, Johannes Hoogeveen, and Alvin Etang. 2014. "Response Behavior and Attrition Management in Mobile Phone Panel Surveys." Unpublished working paper, Twaweza, Dar es Salaam, Tanzania; World Bank, Washington, DC.

Twaweza. 2013. "Sauti za Wananchi: Collecting National Data Using Mobile Phones." Twaweza, Dar es Salaam, Tanzania.

WFP (World Food Programme). 2015. "Post Harvest Improvement in Household Food Security Halts." mVAM Food Security Monitoring Bulletin 7, DR Congo, Mugunga 3 Camp (February), WFP, Rome.

WHO (World Health Organization) and UNICEF (United Nations Children's Find). 2006. "Core Questions on Drinking-Water and Sanitation for Household Surveys." WHO, Geneva.

CHAPTER 6

Data Analysis and Management

Introduction

High-frequency data collection is at the core of a mobile phone panel survey (MPPS) initiative. Nonetheless, once data have been collected, they need to be cleaned and checked before being analyzed and archived. Then, the results must be communicated to stakeholders to help ensure that the data collected are put to use. This chapter discusses four important components that follow data collection: data analysis, report writing, dissemination of the results, and making data available to the public based on the open data principle. The chapter also offers guidelines for the management of MPPS data and examples of dissemination channels.

Data Analysis

Typically, the goal behind an MPPS is not only to produce data for research, but also to provide high-quality data for policy makers, researchers, and other stakeholders. After data collection, data cleaning, and quality checks have been carried out, the next step is to analyze the data (see chapter 1). One cannot jump from a round of mobile phone data collection directly to analysis. First, prior to data analysis, sample weights must be identified and reweighting must be conducted to correct for nonresponses (see chapter 2).

The analysis of survey data may involve two or more units of analysis: individuals and households are often separate units of analysis in household surveys. The unit of analysis used depends on the objective of the analysis and whether analyses are feasible with the available data. Depending on the objective of a survey, once a clean baseline survey dataset has been established, implementers may opt to put together a full report, unpack the baseline to create short module reports, or wait for the completion of the call rounds to present two data points, that is, the baseline and mobile phone interview, for each module.

Because phone interviews are short, analysis can typically be completed quickly. By merging the information collected during the call rounds—the mobile phone interviews—with the more elaborate information collected in the baseline, more meaningful results may be obtained. This permits reporting on an array of issues that can be broken down by subcategories such as rural or urban area, welfare quintiles, sex, age, household size, and so on. The data collected through MPPSs can easily be used to report on a single issue, for example, food security or the effects of weather conditions. However, the methodological advantages of MPPSs may be utilized most effectively by tracking frequent changes. Figure 6.1 shows, for example, trends in food security among various categories of households in Madagascar between March 2014 and January 2015.

These short-term changes could easily be missed if data were collected during only one month during the March 2014–January 2015 time period, which is typically the case in traditional cross-sectional surveys. Reporting on such short-term changes can be revealing because it facilitates regular monitoring and provides information quickly on new or emerging issues such as food insecurity in the Sahel region, ethnic tensions, and health shocks.

If the implementers have distributed mobile phones during the baseline survey, one must be cautious about reporting information that is directly or indirectly associated with mobile phone ownership. Such information can be misleading if it is extrapolated to the entire population given that the mobile phones have been distributed to the households or the households may have already owned the phones, whereas this may not be typical across the population, where many households may not own mobile phones, particularly poor households or households in areas not well served by mobile phone networks. Thus, reporting that 93 percent of the adult population has mobile money accounts, that is, accounts with mobile network operators, may be misleading if household respondents in

Figure 6.1 In the Past Seven Days, Have You Worried Your Household Would Not Have Sufficient Food?

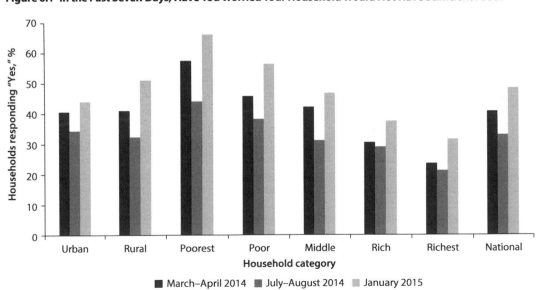

Source: Listening to Madagascar survey data.

Mobile Phone Panel Surveys in Developing Countries • http://dx.doi.org/10.1596/978-1-4648-0904-0

the sample were given the phones that enabled them to open the mobile money accounts, while individuals outside the sample may be more likely not to have such accounts simply because they do not have mobile phones.

Report Writing

While the analysis of data is an inevitable step before the results can be successfully communicated to stakeholders, it is by no means the only step. To ensure that the survey findings gain the attention they deserve and that the statistical results capture an audience's imagination, it is crucial not merely to produce figures and tables, but also to tell stories. In practice, this means compiling survey reports that are accessible and interesting to read and that succeed in bringing complex and diverse findings together into a compelling, coherent narrative. In many cases, the communication and writing skills required to accomplish this may not be found in the data analyst. Nonetheless, the person who prepares the reports must have a sound understanding of the underlying statistical analyses. An important element in the effective communication of results is the use of appealing, meaningful data visualizations. Translating data into graphs that communicate findings in a clear and accessible manner without overwhelming the reader is not an easy task and can require specific creative and technical skills, particularly if data visualization tools are used that are more advanced than Microsoft Excel, such as Infogram or Tableau.

The survey implementer should design the templates for the presentation of the mobile survey results. For each phone survey round, key results can be disseminated in a short report or brief. Results can be reported at different levels depending on the sampling frames. If the survey was nationwide, the results can be reported at nationwide, but also with a distinction between rural and urban areas. Besides presenting and contextualizing descriptive analyses of the main indicators, the report might also offer findings disaggregated by relevant subgroups by sex, age, household size, wealth quintiles constructed according to assets or consumption, and so on. Any statistically meaningful disaggregation of the data depends on the number of observations in the categories of interest (see chapter 2). The approach of the World Bank's Listening to Africa Project and the Sauti za Wananchi survey of Twaweza involves the production of a short report for each mobile phone survey topic for use by journalists, politicians, decisions makers, researchers, and other stakeholders. For example, if a mobile phone survey has collected information on (1) the water sources people usually use, (2) whether these sources were available in the previous week, and (3) the combined waiting and travel time necessary for the collection of water, then a report may be produced presenting information in percent shares on the types of water sources, the availability of these various water sources, and average travel and waiting times in rural and urban areas. In addition, the report might include information on percent changes in these variables since the baseline survey or since earlier mobile phone surveys on the same topic. Information on these variables might also be broken down by the educational attainment of heads of household, household wealth, or the distance of a community to, say, the district center. The information should be presented in an accessible format—graphics, tables, maps, and so on—and may be complemented with information from secondary sources on, for instance, public water expenditures, the relationship between waterborne disease and access to clean water, or changes in water tariffs. As a general rule, the report should be in a language used by the target audience. An example of an MPPS report is provided in appendix F.

Data Dissemination

Once reports and visualizations have been prepared, they need to be communicated in a targeted and planned manner. Any data collection exercise should be based on a clear understanding of the purpose: who is to be reached, and what is to be achieved with the data and the reports that are produced based on the data? While the answers to these questions vary across different MPPS initiatives, a sound communication strategy is always crucial if the reports are to reach targeted groups, catch the attention of the media, and have an impact in public discussions. The expertise of experienced and well-connected communication professionals can make a huge difference in the impact that a project can achieve.

Cleaned and anonymized baseline survey data and mobile phone survey data and the associated reports should be made publicly available through various channels (see below) and stored in the project's archives. In the World Bank Listening to Africa surveys, the aim has been to release the data to the public within four weeks of the completion of the data collection process. This is possible given that phone interviews are short. This means that data from completed mobile phone rounds should be disseminated while the MPPS is still ongoing, that is, publishing the data on completed rounds while phone interviews are still being conducted in current rounds. The dissemination process should be extensive to reach as many stakeholders and as much of the target audience as possible.

Significant resources need to be invested in dissemination, for example, by organizing press conferences, websites, relevant Twitter accounts, stakeholder forums and other sessions bringing together stakeholder individuals, groups, and institutions on the topics under discussion, and other social media platforms. Not all these dissemination channels will be relevant in all cases. The implementers must first carefully select appropriate target groups and stakeholders and then identify the various channels of dissemination available to this target audience. The question is: who can be reached most effectively in what way? Timing is important. For example, if the reports are published online and at a press conference simultaneously, there may be little or no incentive for journalists to attend the press conference. Rather, offer journalists exclusive access to reports at a press conference and then publish online later. The implementers might also consider making an exclusivity deal with a media outlet. They might propose to supply the media outlet regularly with interesting bits of relevant statistical information, such as that 20 percent of households in the country own a bicycle, as long as they publish the bits with appropriate fanfare.

Data and the results of analysis might also be provided regularly as feedback to respondents, who might be notified systematically about the release and use of data or about media reporting on the survey results. Respondents might thus be informed that the key findings of a nutrition survey conducted the previous month were published in the local daily newspaper. An additional benefit of providing such feedback is that it may encourage respondents to participate in future surveys because their voices are obviously being heard. However, one must also be cautious in providing information to respondents because this may lead to response convergence in future surveys. If you inform respondents that the results of the nutrition survey during the previous month suggest that 90 percent of households in the country eat three meals a day, this may encourage the respondents to answer the same question about meals in a future survey in this way so that they are among the more fortunate 90 percent.

The Open Data Principle

The international community is increasingly being spurred to follow open data principles, which are critical to improving access to high-quality data that can be used for a variety of purposes, particularly near real-time decision making. According to open data principles, anonymized data, reports, questionnaires, and other survey documentation should be made available online in a timely, easily accessible manner free of charge. Table 6.1 indicates several platforms on which data and documentation related to MPPSs can be found.

Table 6.1 Platforms Containing Data and Documentation on MPPSs

Survey	Organization	Location	Website
Listening to Africa	World Bank	Washington, DC	http://www.worldbank.org/en/programs/listening-to-africa
	National Institute of Statistics	Antananarivo, Madagascar	http://instat.mg/
	National Statistical Office	Lilongwe, Malawi	http://www.nsomalawi.mw/
	National Agency of Statistics and Demography	Dakar, Senegal	http://www.ansd.sn/
	National Institute of Statistics and Economic and Demographic Studies	Lomé, Togo	http://www.stat-togo.org/
Listening to Dar	Twaweza	Dar es Salaam, Tanzania	http://www.twaweza.org/go/listening-to-dar
Sauti za Wananchi	Twaweza	Dar es Salaam, Tanzania	http://twaweza.org/go/sauti-za-wananchi-english/
Wasemavyo Wazanzibari Mobile Survey, Zanzibar	International Law and Policy Institute	Arusha, Tanzania	http://www.wasemavyowazanzibari.info

CHAPTER 7

How Much Does It Cost?

In the end, whether one carries out a mobile phone panel survey (MPPS) may depend on the cost. If potential implementers have been planning to conduct a panel survey and become familiar with the literature on standard face-to-face panels, they will have read that these tend to be expensive. This is mainly because the cost of tracking and then revisiting individuals or households who split off from the original sample again and again can quickly add up. So, is an MPPS any different? The answer depends on several factors.

The cost of an MPPS depends, for example, on the sample size, the duration of the survey, the choice of a baseline survey, the expertise hired, the geographical coverage of the survey, the cost of making a call and the wage rates in a country, and the amounts of any incentives paid to households. The geographical coverage of the survey matters in terms of the baseline survey only, whereas the other issues are relevant for the phone survey rounds and, possibly, the baseline, face-to-face survey. The incentive payment is typically small, no more than $1.00 per successful interview (see chapter 5). The cost of a call varies from country to country and is lower where there is more competition and greater regulation of service providers.

This chapter offers an overview of key budget items for an MPPS project. Appendix G offers a detailed, but by no means exhaustive checklist of items that typically enter the costing exercise. The biggest driver of costs by far is the size of the sample, which, in turn, depends on the objective of the survey being conducted. A detailed discussion of sample size is provided in chapter 2. It might be as low as 500 respondents to several thousand. The surveys that inform the lessons presented in this book have sample sizes of between 500 and 2,000 respondents.

Once a sample size is chosen, there are three big costs in conducting an MPPS. The first major expense is the baseline survey. The detailed description in chapter 3 on what needs to be done to carry out a successful baseline survey provides an indication of how the costs arise. Nonetheless, potential implementers of an MPPS might consider realizing a baseline survey, which is a key feature of a successful MPPS. Some of the costs of a baseline survey are unavoidable.

However, depending on the survey's timing and with careful planning to prepare for the survey, it may be possible to reduce some of the costs. For instance, perhaps a nationally representative household survey has recently been completed, and most of the information one would need from a baseline has already been collected. In that case, one might select an appropriate subsample of the household survey, revisit the respondents to hand over mobile phones and, if necessary, solar chargers and explain the operation of the mobile phone survey rounds. This would reduce the time the enumerators need to spend with households, thereby cutting some of the costs. There might also be some marginal savings from using tablets rather than paper, depending on the cost of printing.

The second major expense is the hardware required to execute a successful MPPS. These typically include the phones that are distributed to respondent households and possibly solar chargers, especially in rural areas, to limit the likelihood of nonresponse. How costly this step is depends on the type of phones and chargers distributed. All the surveys used as examples in this handbook are based on the cost of low-end phones and small solar chargers. Although the costs of these two items are modest, usually about $40 per respondent, they quickly add up depending on the number of people to whom they are given. If the mobile phone survey is being conducted in an area in which most people have phones and the power supply is not an issue, then this cost may be negligible, although it may still be advisable to consider providing other incentives for participation. In many low-income countries, supplying phones is unavoidable. All the surveys cited in this handbook have been conducted in low-income countries in which phones have been given to all respondents.

The third large source of spending is the call center. This includes personnel—wages for supervisors and interviewers—and hardware costs, such as computers to enter and archive data and phones to call respondents. Most of the expenditure of this part of the budget arises from the fact that the phone interview rounds are repeated over several months or years. The costs depend on how frequently the interviews are conducted. High-frequency contacts, say, weekly over a period of months, can be costly. Conversely, long intervals between calls, say, contacting households once every two or three months, can reduce the costs. However, budget choices also have implications for sample quality in terms of attrition. Both high-frequency contacts and long lags between calls could generate high attrition rates, though the former has the added disadvantage of higher cost. The examples of budgets provided below are therefore based on one call a month for each survey round.

Table 7.1 illustrates the approximate cost of conducting an MPPS in four countries. In all cases, a baseline survey was carried out that was almost identical in terms of design across the four countries. All respondents received a phone, and all rural respondents also received a solar charger. The budgets also assume that the MPPS rounds are run for two years (24 months) among a sample of between 1,500 and 2,000 respondents. The survey in Togo is realized only in Lomé, the capital and main city. The other three surveys are nationally representative.

Table 7.1 Cost of Conducting a Mobile Phone Panel Survey, Four Sample Countries
U.S. dollars

Parameter	Malawi	Senegal	Madagascar	Togo
Survey start date	August 2014	November 2014	April 2014	March 2014
Baseline	180,614	64,901	108,010	57,825
Establish call center and purchase equipment (phones, solar chargers, SIM cards)	106,120	185,875	28,636	19,780
Running the call center, per phone round	8,767	7,440	11,344	2,197
Sample size, respondents	1,504	1,500	2,000	500

Table 7.2 Cost Comparison, an MPPS and an LSMS Complex Multitopic Survey, Malawi
U.S. dollars

Indicator	Third Integrated Household Survey, Integrated Household Panel Survey	Listening to Malawi, baseline	Listening to Malawi, phone survey	Listening to Malawi, phone survey w/o a call center, phones, or chargers
Total cost per survey	600,000	180,614	13,188	8,767
Sample size, number of households	4,000	1,504	1,504	1,504
Cost per household	150.00	120.00	8.80	5.80
Number of questions	2,863	923	42	42
Cost per question	0.06	0.13	0.20	0.14

The overall budgets of the nationally representative surveys total around $408,000 in Madagascar and $500,000 in Malawi for a baseline survey, plus 24 monthly phone survey rounds, while the Lomé survey cost around $100,000. Table 7.1 shows the cost range, which varies by country. The cost of a call is high in Malawi compared with the other countries.

It is tempting to compare these costs with the cost of a standard face-to-face survey, but there are no obvious and easy ways to accomplish this given the differences in sample sizes, the frequency of data collection in an MPPS, the complexity of questions in traditional surveys, and the number of questions per module. Nonetheless, a quick, back-of-the-envelope calculation shows that traditional surveys that analyze thousands of responses cost less per question. For example, a typical complex, multitopic household survey that is in the field for a year might cost around $140–$150 per household—excluding technical assistance in sampling and data entry—and collect data on responses to roughly 3,000 questions or about $0.06 per question, compared with $0.20 per question in a mobile phone survey (table 7.2). However, if we compare the annual cost of running an MPPS with the corresponding cost of a typical light face-to-face survey that collects modules similar to the 12 modules of a mobile phone survey or equivalent surveys, then an MPPS may be cheaper. Examples of such light face-to-face surveys include welfare monitoring surveys, core welfare indicator questionnaire surveys, and so on. Assuming that the cost per household in fielding a light face-to-face survey is not much different from the cost of fielding a Living Standards Measurement Study (LSMS) survey, then an MPPS becomes a less expensive alternative. For the same number of households, say, 2,000, a light face-to-face survey would cost around $300,000 a year, while the MPPS would cost roughly $150,000 a year.

While these back-of-the-envelope cost comparisons are useful, they should be treated with caution. First, these mobile phone surveys have modest aims. They are suitable for learning quickly about service failures and for monitoring the pulse of the population on a frequent basis, especially if there is a crisis brewing. Second, the comparison is also inappropriate in the sense that these mobile phone surveys are not intended to replace standard surveys, but rather as complements. Third, cost considerations must be balanced with the other attractive intangibles of various types of surveys. For example, it is difficult to cost the flexibility inherent in the ability to introduce new questions on short notice, the rapid turnaround in data entry and analysis, and the high-frequency panels of a phone survey.

Head of Household Consent Form

Introduction

Good Morning/Good Afternoon/Good Evening. My name is ....
from (INSERT NAME OF DATA COLLECTION FIRM/INSTITUTION), an
independent market and social research firm based in Dar es Salaam. Our offices
are located in (PROVIDE OFFICE LOCATION DETAILS).

(DATA COLLECTION FIRM) is partnering with (NAME OF PARTNER)
(PROVIDE DETAILS OF THE MISSION OF THE PARTNER) to carry out a
mobile phone survey known as "INSERT NAME OF SURVEY."

Your household has been randomly selected to participate in this mobile
phone survey. I want to invite one of the members of your household to take part
in this survey. First, I will talk to you about the study and answer any questions
that you may have.

Purpose of the Research Project

"INSERT NAME OF SURVEY" is a nationwide/regionwide mobile phone panel
survey in (INSERT COUNTRY/CITY/ZONE). The survey has the potential to
improve the availability of information and contribute to the development of
better public services and life in (INSERT COUNTRY/CITY/ZONE). The study
will interview (INSERT THE NUMBER OF RESPONDENTS) respondents.

"INSERT NAME OF SURVEY" is divided into two phases. The first phase is
known as the baseline survey in which we seek to collect data by having a face-
to-face interview. This is what we are doing today. The second phase of "INSERT
NAME OF SURVEY" is a mobile phone survey. The mobile survey will be a
short interview of about 10 to 15 minutes. It will be carried out once every
(INSERT THE FREQUECNY OF THE SURVEY) for (INSERT THE PROJECT
TIMELINE).

Procedures

If you agree to allow me to interview one household member, I will randomly select from all eligible household members ages 18 years or older. Random selection means selecting by "lucky draw": each adult in the household has the same chance of being selected. After selection of the main respondent, we need to respect that choice.

I will ask him/her questions about (PROVIDE THE LIST OF THEMATIC AREAS AND ANY OTHER PEOPLE TO BE INTERVIEWED DURING THE BASELINE, FOR EXAMPLE, CHILDREN AND EXPECTANT WOMEN). If a respondent is not able to answer any question, he/she will ask other household members who are able to respond.

To facilitate round 2 of this survey, we shall give the selected respondent a mobile phone so that we can reach him or her. In the event the respondent does not want to receive the phone and would like to be interviewed using his/her own phone and number, this is okay with us. We also appreciate that lack of a charging facility is a major problem in developing countries such as ours. We shall provide a solar charger that may be shared with another participant. The two items belong to the survey, and ownership will only be transferred to the household member at the end of the research period, that is, after (INSERT PROJECT TIMELINE). During the research period, the phone and charger should be available for the main respondent of the survey, but they may also be used for communication by other household members.

Confidentiality Clause

The family member's responses or information about this household will not be shared with anyone outside the study team. The reports might say, for example, "80 percent of Tanzanians are not connected to the national electric grid." No personal information will ever be shared.

Benefits

There is no direct benefit to you or the person we select to participate. Neither you nor the person will receive compensation. However, the survey has the potential to improve the availability of information and contribute to the development of better public services and life in our country. The phone and charger that we shall leave in the household can be used for the communication needs of members of the household.

Risks of Participation

There is no risk of participation in this survey.

Who Do I Call If I Have a Question?

If you have a question or concern about this survey, please contact (INSERT NAMES AND CONTACT DETAILS OF 2 KEY TEAM MEMBERS) on Monday–Friday from 8:00 am to 5:00 pm.

I have given you highlights of the "INSERT NAME OF SURVEY." Do you agree to allow your family member to participate in this survey?

Yes	1	THANK THE RESPONDENT AND CONTINUE
No	2	ESTABLISH THE REASON FOR REFUSAL AND ADDRESS IT. IF HE/SHE REFUSES, THANK HIM/HER AND RECORD THIS OUTCOME. CHOOSE ANOTHER HOUSEHOLD ACCORDING TO INSTRUCTIONS.

"I have read the consent form completely before the study participant, and the study participant voluntarily agreed to allow the family members to participate in the study."

_____   _____   _____
Signature of the enumerator   **Name of enumerator**   **Date**

PROVIDE A COPY OF THE CONSENT FORM TO THE HEAD OF HOUSEHOLD

Respondent Agreement Form

This agreement is made on (**INSERT DATE OF SIGNING THE AGREEMENT**)
.... between (**INSERT NAME OF IMPLEMENTERS**),
(**INSERT ADDRESS OF IMPLEMENTERS**), of the one party and (**INSERT
NAME OF RESPONDENT**) ....
"Agreeing party" of the other party.
(**INSERT NAME OF IMPLEMENTERS**) .... and
(**INSERT NAME OF RESPONDENT**) ....
agree to the following:

1. (**INSERT NAME OF IMPLEMENTERS**)...
   will hand you a/the following; (**INSERT THE NAME AND QUALITY OF
   DEVICES**).

2. The devices/gadgets are given to you to facilitate data collection in the
   (**INSERT NAME OF MPPS PROJECT**), a study that seeks to collect data
   using mobile phones across (**INSERT TARGET GEOGRAPHICAL AREA**) to
   (**INSERT PURPOSE OF THE MOBILE PHONE PANEL SURVEY**).

3. You have voluntarily accepted to participate in (**INSERT NAME OF MPPS
   PROJECT**).

4. The devices/gadgets are provided to you to facilitate data collection. In the
   event you desire to drop out of the survey before the end of the survey,
   you will return the mobile phone handset to the selected survey facilitator in
   your village/street. (**ONLY INCLUDE ARTICLE 4 OF THE CONTRACT
   IF YOU INTEND TO TAKE THE MOBILE PHONE FROM THE
   RESPONDENTS BEFORE THE END OF THE SURVEY AND IF YOU
   HAVE ORGANIZED THE RESPONDENTS IN A GROUP OR ASSIGNED
   A SURVEY FACILITATOR DURING THE BASELINE.**)

5. Lost mobile phone handsets and accessories will not be replaced, and, in case
   your handset is stolen or misplaced, you will report the incident to the survey
   facilitator as well as the village/street authorities.

6. Every time a person is interviewed in **(INSERT NAME OF MPPS PROJECT)**, **(INSERT NAME OF IMPLEMENTERS)** will send you an airtime recharge credit or mobile money cash transfer of .... **(INSERT AMOUNT ACCORDING TO THE LOCATION: THE AIRTIME RECHARGE CREDIT IS THE RESPONDENT INCENTIVE.)**

7. The mobile phone handset and its accessories will belong to you at the end of the study, that is, **(INSERT PROJECT PERIOD)** from the date on which we will have completed baseline data collection across the country, which is **(INSERT PROJECT COMPLETION DATE)**.

We, the undersigned, fully understand and agree to the above.

For: (INSERT NAME OF IMPLEMENTERS)

Name: **(INSERT NAME OF THE PROJECT LEAD)** Position: **(INSERT THE POSITION OF THE PROJECT LEAD)**

Signature: Location:

Date:

For: Agreeing Party

Name: **(INSERT NAME OF THE RESPONDENT)**

Signature: Location:

Date:

Baseline Survey Checklist

Mobile Phone Panel Survey – Baseline Survey (Checklist)

No.	Activity	Responsible party	Status	Action	Date of completion
A. Research design					
1	Research design				
2	Sampling plan and enumeration area (EA) maps				
3	Research clearance				
B. Technological considerations					
4	Selection of hardware: phone and solar chargers				
5	Purchase of hardware: phone and solar chargers				
6	Registration of SIM cards				
C. Development of survey instruments					
7	Community questionnaire				
8	Listing form				
9	Head of household consent form				
10	Baseline survey questionnaire				
11	Respondent agreement form				
12	Community materials				
13	Respondent training manual				
14	Enumerator training manual				
15	Field update form				
D. Pretest					
16	Pretest team				
17	Pretest training				
18	Conducting the pretest				
19	Pretest debriefing				
20	Revision of the survey instruments				
E. Pilot phase					
21	Pilot team				
22	Pilot training				

table continues next page

Mobile Phone Panel Survey – Baseline Survey (Checklist) *(continued)*

No.	Activity	Responsible party	Status	Action	Date of completion
23	Pilot implementation				
24	Pilot debriefing				
25	Pilot data checks				
26	Revision of the survey instruments and implementation plans				
F. Fieldwork					
27	Recruitment of field team				
28	Selection of field team				
29	Training the field teams				
30	Training the baseline team				
31	Fieldwork budgets				
32	Field logistics plan				
33	Recruiting the data entry team (ONLY FOR Paper-and-Pencil Interviewing [PAPI])				
34	Training the data entry team				
G. Call center preparations					
35	Setting up the call center: hardware and location				
36	Recruiting call center team				
37	Training call center team				

Sample Mobile Phone Questionnaire

Listening to Africa, Nutrition and Food Security Module

Today, we would like to ask you about food consumption in your household.

A. NUTRITION

Item		A1. *In the past <u>one week</u> (seven days), how many days have you or others in your household consumed any of the following?* **IF NOT CONSUMED, RECORD ZERO**
		NUMBER OF DAYS
A.	**Cereals, grains, and cereal products** (maize grain/flour, green maize, rice, finger millet, pearl millet, sorghum, wheat flour, bread, pasta, other cereal)	
B.	**Roots, tubers, and plantains** (cassava tuber/ flour, sweet potato, Irish potato, yam, other tuber/plantain)	
C.	**Nuts and pulses** (bean, pigeon pea, macademia nut, groundnut, ground bean, cow pea, other nut/pulse)	
D.	**Vegetables** (onion, cabbage, wild green leaves, tomato, cucumber, other vegetables/leaves)	
E.	**Meat, fish, and animal products** (egg, dried/ fresh/smoked fish, excluding fish sauce/powder; beef; goat meat; pork; poultry; other meat)	

F.	**Fruits** (mango, banana, citrus, pineapple, papaya, guava, avocado, apple, other fruit)	
G.	**Cooked foods from vendors** (boiled or roasted maize, chips, boiled cassava, boiled eggs, chicken, meat, fish, doughnuts, samosa, meal eaten at restaurant, other cooked foods from vendors)	
H.	**Milk and milk products** (fresh/powdered/soured milk; yogurt; cheese; other milk products, excluding margarine/butter or small amounts of milk for tea/coffee)	
I.	**Fats/oil** (cooking oil, butter, margarine, other fat/oil)	
J.	**Sugar/sugar products/honey** (sugar, sugarcane, honey, jam, jelly, sweets/candy/chocolate, other sugar products)	
K.	**Spices/condiments** (salt, spices, yeast/baking powder, tomato/hot sauce, fish powder/sauce, other condiments)	
L.	**Beverages** (tea; coffee; cocoa; millo; squash; fruit juice; freezes/flavored ice; soft drinks such as Coca-Cola, Fanta, Sprite, and so on; commercial traditional-style beer; bottled water; bottled/canned beer; traditional beer; wine or commercial liquor; locally brewed liquor; other beverages)	

B. FOOD SECURITY

B1. In the past seven days, have you worried that your household would not have enough food? Answer: _____ 1 = Yes; 2 = No.

B2. In the past seven days, how many days have you or someone in your household had to… **IF NO DAYS, RECORD ZERO**	DAYS
a.	Rely on less preferred or less expensive foods?	
b.	Limit portion size at meal times?	
c.	Reduce the number of meals eaten in a day?	
d.	Restrict consumption by adults so small children could eat?	
e.	Borrowed food or relied on help from a friend or relative?	

B3. How many meals, including breakfast, are taken per day in your household?		NUMBER
a.	Adults	
b.	Children (6–59 months) **LEAVE BLANK IF NO CHILDREN**	

B4. In the past six months [**instead of six, insert the number of months since the last survey on this topic**], have you been faced with a situation that you have not had enough food to feed the household? Answer: _____ 1 = Yes; 2 = No >>**B7**

B5. When did you experience this incident in the last six months [**instead of six, insert the number of months since the last survey on this topic**]?

MARK X IN EACH MONTH OF 2013 AND 2014 THE HOUSEHOLD DID NOT HAVE ENOUGH FOOD

LEAVE CELL BLANK FOR FUTURE MONTHS FROM INTERVIEW DATE OR MONTHS MORE THAN SIX MONTHS AGO FROM INTERVIEW DATE [**number of months since the last survey on this topic**].

2013											
Jan	Feb	Mar	Apr	May	June	July	Aug	Sep	Oct	Nov	Dec

2014											
Jan	Feb	Mar	Apr	May	June	July	Aug	Sep	Oct	Nov	Dec

B6. What was the cause of this situation? LIST UP TO THREE *[Do not read options. Code from response]*.

CODES FOR B6:

CAUSE 1	CAUSE 2	CAUSE 3

1 = Inadequate household stocks because of drought/poor rains

2 = Inadequate household food stocks because of crop pest damage

3 = Inadequate household food stocks because of small land size

4 = Inadequate household food stocks because of lack of farm inputs

5 = Food in the market was expensive

6 = Unable to reach the market because of high transportation costs

7 = No food in the market

8 = Floods/water logging

9 = Other (Specify): _____

B7. Does your household cope with food shortages in any of the following ways?		1 = Yes 2 = No
A.	Reduce number of meals eaten in a day	
B.	Limit portion size at meal times	
C.	Rely on less preferred or less expensive foods	
D.	Change food preparation	
E.	Borrow money or food or rely on help from a friend or relative	
F.	Postpone buying tea/coffee or other household items	
G.	Postpone paying for education (fees, books, and so on)	
H.	Sell household property, livestock, and so on	

B8. In case of food shortage, who eats less? Answer: _____
1 = Boys 0–15 years
2 = Girls 0–15 years
3 = Boys and girls 0–15 years
4 = Men 16–65 years
5 = Women 16–65 years
6 = Men and women 16–65 years
7 = People over 65 years old
8 = Everyone eats equal amounts

Sample Phone Survey Round Calendar

Activities	Current month, M, by week, W			
	W1	W2	W3	W4
Ensure the monthly survey of the current month M	■	■	■	■
Prepare the survey for month M + 1				
Identify the theme of the month and the expert of the month M + 1	■			
Prepare a draft questionnaire for month M + 1 with the expert of the month	■			
Validate the questionnaire with the steering committee during a meeting			■	
Finalize the questionnaire according to the latest observations of the committee			■	
Transfer the questionnaire to the call center				■
Provide data control schema related to the questionnaire to the call center				■
Train the team of the call center on the new concepts of the questionnaire				■
The call center programs the questionnaire and data control schema in its computer-assisted telephone interviewing (CATI) system				■
Analyze data from the M − 1 survey; prepare the report; and distribute it				
Recover the data basis of the M − 1 survey		■		
Retrieve the collection report of the monthly survey M − 1		■		
Retrieve the list of households that have not been interviewed in the month M − 1		■		
Treat the data basis of the survey M − 1		■		
Draft the simplified tables and the four-page report of month M − 1 with the relevant expert		■		
Validate the four-page report with the steering committee			■	
Prepare and distribute the report of the M − 1 survey				■
Prepare the data and put them on the website				■
Follow up with nonrespondent households in the field, if necessary	■	■	■	
Hold monthly meeting of the steering committee		■		

Sauti za Wananchi Report 9 (March 2014)

Introduction

Open the newspaper on any given day, and you will find stories related to violence and theft. For example, *The Citizen* newspaper's headlines on 7 January 2014 included "Police hunt for robbers who killed 2 'bodaboda' [motorcycle taxi] operators;" "Chadema supporters clash in Dar;" and "One dead, police move in as mob goes on rampage and loots mine." However, newspapers are said to prefer sensationalism, seeking out incidents with high headline value rather than the balanced truth. So, how likely is any such headline event to affect the life of the average Tanzanian? How often have people actually been victims or witnesses of crime? And what is their recourse when crimes are committed? Do people believe that they can rely on the justice system?

Sauti za Wananchi, Africa's first nationally representative mobile phone survey, seeks answers to these and other questions around security in Mainland Tanzania. This brief presents nine facts on security in Tanzania using the most recent data available. The findings are based on the eighth round of Sauti za Wananchi (www.twaweza.org/sauti). Calls were made between 3 and 17 October 2013; data include responses from 1,662 households. This brief also presents findings from the Afrobarometer surveys and the Sauti za Wananchi baseline survey, which was implemented between October and December 2012.

The key findings, based on respondent reports, are

- 49 percent of Tanzanians have never had anything stolen, but 20 percent had something stolen in the last six months.
- 46 percent of Tanzanians recently observed violence in public.
- In more than half of all villages (or streets) in Tanzania, no police officer is posted.

This a reproduction of "Are We Safe?: Citizens Report on the Country's State of Security," Sauti za Wananchi Brief 9 (March 2014), Twaweza, Dar es Salaam, Tanzania, http://www.twaweza.org/uploads/files/SzW-R8-Security280214FINAL-EN.pdf. Used with permission.

- Tanzanians are not aware of the police and fire emergency numbers.
- Corruption and slow response are the main barriers to reporting crime to the police.
- 45 percent of Tanzanians felt unsafe walking in their neighborhood at least once in the last year.

Nine Facts about Security in Tanzania

Fact 1: 20 Percent of Tanzanians Report Cases of Theft in the Last Six Months

For sustainable economic growth the security of a person's or a community's money and property is crucial: the better the security, the lower the costs of protection, and the more attractive it is to invest. As figure F.1 illustrates, one in two Tanzanians have never had anything stolen. However, one out of five had something of value stolen in the last six months.

The Afrobarometer survey (www.afrobarometer.org) can be used to put these numbers in long-term perspective; see figure F.2. Afrobarometer asks whether anyone in the respondent's family had anything stolen from their home during the past year. The data suggest an upward trend in theft in Tanzania since 2005. The Afrobarometer data also show that theft in Tanzania is currently higher than the average for Africa as a whole (36 percent versus 26 percent).

Fact 2: 46 Percent of Tanzanians Recently Observed Violence in Public

Almost half of Tanzanians (46 percent) report having observed violence in public within the past six months. [See figure F.3.]

Only small minorities are also aware of security incidents across the country reported in the media.

Respondents reported to be aware of the following incidents: Arusha church bombings (May 2013, 17 percent), Mtwara gas riots (June 2013, 13 percent),

Figure F.1 When Was the Last Time Money or Anything of Value Was Stolen from You?

Source: Sauti za Wananchi mobile phone survey, round 8, October 2013.

Figure F.2 Experienced Theft in the Home

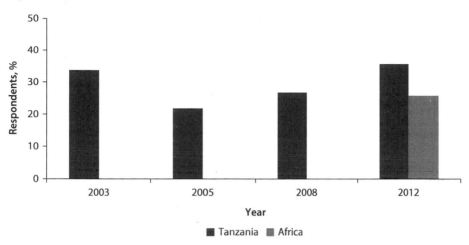

Source: Afrobarometer surveys.

Figure F.3 When Was the Last Time You or Your Household Member Observed Violence in Public?
percent

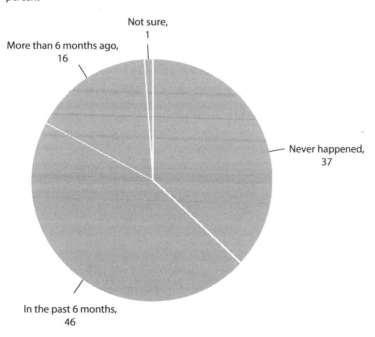

Source: Sauti za Wananchi mobile phone survey, round 8, October 2013.

Chadema rally bombings in Arusha (June 2013, 11 percent), and Zanzibar acid attacks on British volunteers (August 2013, 8 percent).

Fact 3: More Than Half of Communities in Tanzania Do Not Have Police Officers

According to interviews with the village executive officers and urban neighborhood ("street") chairpersons, 62 percent of communities in Tanzania do not have a designated police officer [see figure F.4]. The situation is worse in rural areas, where 76 percent of the villages report not having a police officer posted there.

Fact 4: Tanzanians Are Not Aware of the Police and Fire Emergency Numbers

Crime and fires occur unexpectedly, but, once they occur, the police and firefighters, respectively, are often better placed to control the situation than ordinary citizens. Mobile telephony has grown over time, so many Tanzanians should have access to the means to contact the police and fire services. However, only 15 percent of citizens know the police emergency number, while only 6 percent know the fire emergency number [see figure F.5]. The survey did not check the functionality of these emergency numbers.

Fact 5: In Case of Crime, 47 Percent of Tanzanians Turn to Police First

Asked who they would turn to in case they were victim of a crime, most Tanzanians indicate the police and local security organizations [see figure F.6]. A majority of urban residents mention the police. Although rural citizens also

Figure F.4 Number of Police Officers in a Village/Street

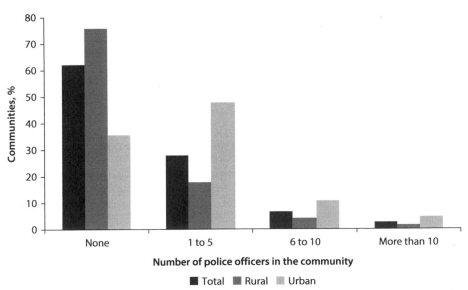

Source: Sauti za Wananchi, Baseline Survey, Community, October–December 2012.

Figure F.5 Do You Know the Police and Fire Emergency Numbers?

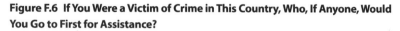

Source: Sauti za Wananchi mobile phone survey, round 8, October 2013.

Figure F.6 If You Were a Victim of Crime in This Country, Who, If Anyone, Would You Go to First for Assistance?

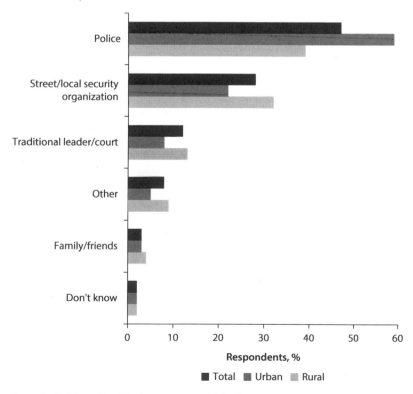

Source: Sauti za Wananchi mobile phone survey, round 8, October 2013.

Mobile Phone Panel Surveys in Developing Countries • http://dx.doi.org/10.1596/978-1-4648-0904-0

largely turn to the police, similar numbers report turning to local security organizations. Few indicate not knowing who to turn to.

Fact 6: Corruption and Slow Response Are Barriers to Reporting Crime

Sauti za Wananchi respondents were asked to indicate what barriers, if any, they
think citizens might encounter in considering to report crimes to the police.

The respondents outlined the following as the two main issues: police ask for
bribes (22 percent), and police don't listen or care (22 percent). [See figure F.7.]
On the other hand, 16 percent of respondents mentioned that people do report
crimes to the police. We note that these numbers differ somewhat from the
Afrobarometer 2012 results, where distance to a police station is mentioned as
the main obstacle.

Fact 7: Killings Most Frequently Attributed to Mobs

When asked whether they have ever heard of anyone in their neighborhood
being threatened, beaten, or stoned by community police, an ordinary citizen, a
mob, the police, or the army, 54 percent say yes [see figure F.8]. In addition,
31 percent report having heard that someone in their neighborhood was killed by
one of these groups. In case someone was killed, this was reportedly most often
(19 percent) done by a mob. When someone was threatened, beaten, or stoned,
this was reportedly done by community policy in most cases (31 percent).

Note again that these numbers are not based on first-hand observations, but are
incidents that were "ever heard" by respondents. Taken at face value, these data
mean that a Tanzanian, on average, is much more likely to be threatened, beaten,
or killed by a fellow citizen or group of citizens than by the police or army.

**Figure F.7 What Do You Think Is the Main Reason Many People Do Not Report Crimes Such as Thefts or
Attacks to the Police When They Occur?**

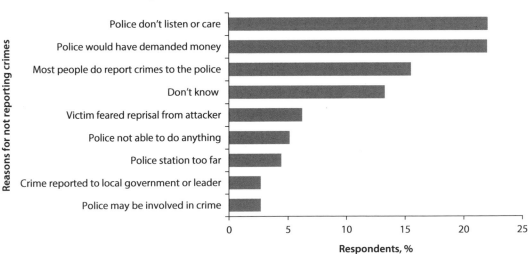

Source: Sauti za Wananchi mobile phone survey, round 8, October 2013.

Figure F.8 Who Perpetrates Violence?

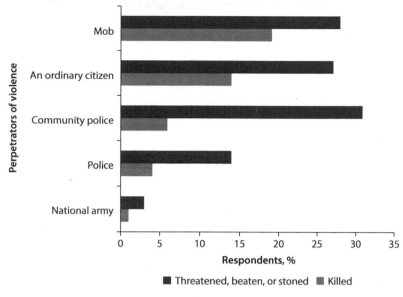

Source: Sauti za Wananchi mobile phone survey, round 8, October 2013.

Citizens were also asked to provide their understanding of the context in which these events took place. When considering killings by a mob and by an individual citizen, the reported underlying issue in most cases was stealing (in 74 percent of the mob cases, and 47 percent of the individual cases).

Fact 8: Most Tanzanians Do Not Believe in the Justice System
Do Tanzanians believe that a person who has committed a crime will be punished according to the law? Figure F.9 shows, first, that respondents have little faith that *any* person will be punished according to the law. Second, there is a view that rich and powerful people are more likely to escape punishment than ordinary citizens.

Fact 9: Many Tanzanians Are Worried about Safety in Their Communities
Overall, 39 percent of the respondents reported fearing a crime in their home; 45 percent mention that they felt unsafe walking in their neighborhoods at least once [see figure F.10]. This resonates with Afrobarometer data that show 43 percent of Tanzanians fear crime in their homes at least once. The Afrobarometer data further show that this statistic for Tanzania is high compared with the average in Africa, of 32 percent.

It appears that citizens are more fearful during election time. More than half (57 percent) of the respondents expressed some fear of becoming a victim of political violence, for example, compared with the 45 percent who felt unsafe walking in their neighborhoods at least once over the last year. [See figure F.11.]

Mobile Phone Panel Surveys in Developing Countries • http://dx.doi.org/10.1596/978-1-4648-0904-0

Figure F.9 If the Following Person Steals or Commits Any Other Crime, Will He or She Be Punished According to the Law?

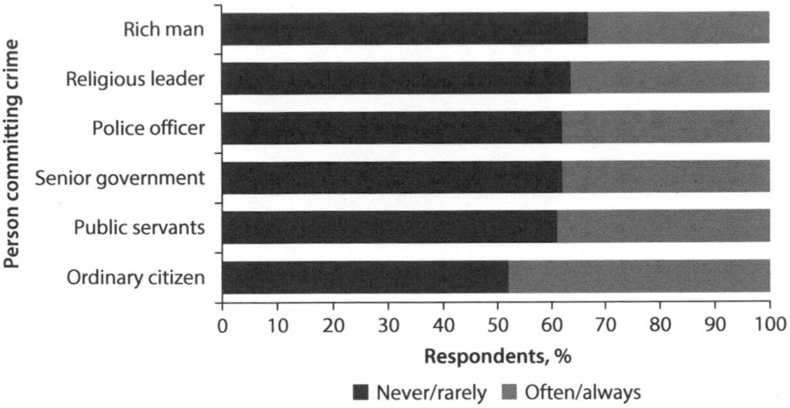

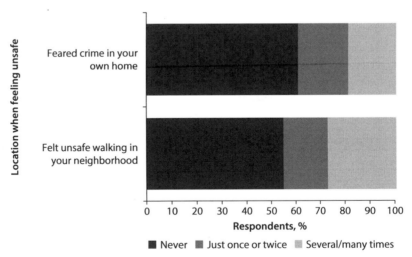

Source: Sauti za Wananchi mobile phone survey, round 8, October 2013.

Figure F.10 Over the Past Year, How Often, If Ever, Have You Been Afraid or Felt Unsafe?

Source: Sauti za Wananchi mobile phone survey, round 8, October 2013.

Conclusion

This brief reports on security as experienced by citizens of Mainland Tanzania. The brief finds that—as in many other societies—Tanzanians experience crime and fear: money and other things of value are sometimes stolen from citizens; and they observe and hear stories of violence in their neighborhoods.

Since the state has a monopoly on the use of force, citizens are supposed to report such incidents to the police who are charged with the responsibility of

**Figure F.11 During the Last General Election, Did You Fear Becoming a
Victim of Political Violence?**
Respondents, %

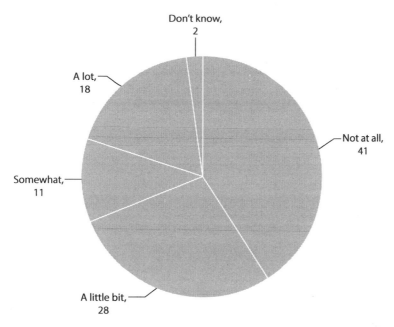

Source: Sauti za Wananchi mobile phone survey, round 8, October 2013.

keeping peace, enforcing the law, protecting people and their property, as well as investigating crimes. For justice according to the law to be done after a crime has been committed, the perpetrator needs to be correctly identified, apprehended, and then convicted by a court of law.

Citizens in Tanzania look to the police for assistance, but access to the formal security system is not easy, particularly in rural areas. In many villages, there are no police officers. In general, police officers are not always perceived as helpful and people typically do not know the police (or fire brigade) emergency phone numbers.

As a result, people often resort to using nonformal ways to deal with those identified as criminals. It is not uncommon for Tanzanians to have heard of someone in their neighborhood being killed by a mob. These incidents often follow a case of theft. Another side of this coin is the fact that many Tanzanians do not believe that criminals will be punished according to the law, particularly if they are wealthy or in a position of authority.

With confidence in the justice system at a low point, and nonformal means of justice open to abuse, the country faces a serious challenge to ensure public security. While expanding formal means of security and resourcing the sector adequately may be part of the solution, the situation may need more creative, out-of-the-box thinking. This may include expanding community policing, which emphasizes closer collaboration between the police and citizens; better regulation of traditional security groups; and promoting ways in which core structures of local governance (for example, village councils, school committees) can be made more responsive to citizen voices and demands. These issues deserve greater public debate, including in the constitutional assembly process.

Checklist of Budget Items for MPPSs

Budget items, MPPS baseline	
Training field enumerators	
• Salary of trainers	
• Per diems, trainees	
• Location	
• Refreshments	
• Printing costs, training materials	
• Stationery	
Fieldwork (pilot baseline and baseline)	
• Salary staff (enumerators, supervisors, drivers)	
• Per diem and accommodation, field teams	
• Field vehicles	
• Fuel for field vehicles	
• Phones to be handed out to respondents	
• SIM cards and call credit for respondents	
• Solar chargers to be handed out to respondents	
• Communication costs, field teams	
• Printing costs, questionnaires	Only for PAPI
• Printing costs, listing forms and maps	
• Printing costs, enumerator/supervisor manuals	
• Mobile devices for data entry (laptops/tablets/mobile phones)	Only for CAPI
• Solar chargers for data entry devices	Only for CAPI
• Data entry software for mobile devices	Only for CAPI
• Stationery	
Analysis and dissemination	
• Data entry	
• Data analysis software	
• Data checking, cleaning, and labeling	
• Data analysis	Only for CAPI
• Report writing	
• Development and upkeep of website	
• Stakeholder events/press conference	

table continues next page

Appendix G *(continued)*

Budget items, MPPS follow-up rounds

Training call center interviewers
- Salary, trainers
- Per diems, trainees
- Location and refreshments
- Printing costs, training materials
- Stationery

Call center infrastructure
- GSM gateway
- Infrastructure for broadband connectivity to service provider
- Furniture (chairs, tables, separators)
- Phones for call center interviewers
- Headsets for call center interviewers
- Computers/tablets for data entry
- Data entry software
- Call center software

Operational costs
- Salary, call center staff: supervisors and interviewers
- Communication costs for interviews
- Credit to transfer to respondents
- Data checking, cleaning, and labeling
- Data analysis
- Report writing
- Stakeholder events/press conferences

Note: CAPI = computer-assisted personal interviewing. GSM = Global System for Mobile Communications. PAPI = paper-and-pencil interviewing.